T0185967

# SpringerBriefs in Electrical and Computer Engineering

**Series Editors**

Woon-Seng Gan, School of Electrical and Electronic Engineering, Nanyang Technological University, Singapore, Singapore

C.-C. Jay Kuo, University of Southern California, Los Angeles, CA, USA

Thomas Fang Zheng, Research Institute of Information Technology, Tsinghua University, Beijing, China

Mauro Barni, Department of Information Engineering and Mathematics, University of Siena, Siena, Italy

SpringerBriefs present concise summaries of cutting-edge research and practical applications across a wide spectrum of fields. Featuring compact volumes of 50 to 125 pages, the series covers a range of content from professional to academic. Typical topics might include: timely report of state-of-the art analytical techniques, a bridge between new research results, as published in journal articles, and a contextual literature review, a snapshot of a hot or emerging topic, an in-depth case study or clinical example and a presentation of core concepts that students must understand in order to make independent contributions.

More information about this series at http://www.springer.com/series/10059

Christoph Guger · Brendan Z. Allison ·
Michael Tangermann
Editors

# Brain-Computer Interface Research

A State-of-the-Art Summary 9

 Springer

*Editors*
Christoph Guger
g.tec medical engineering GmbH
Schiedlberg, Oberösterreich, Austria

Brendan Z. Allison
Department of Cognitive Science
University of California San Diego
San Diego, CA, USA

Michael Tangermann
Brain State Decoding Lab
University of Freiburg
Freiburg, Baden-Württemberg, Germany

ISSN 2191-8112          ISSN 2191-8120 (electronic)
SpringerBriefs in Electrical and Computer Engineering
ISBN 978-3-030-60459-2          ISBN 978-3-030-60460-8 (eBook)
https://doi.org/10.1007/978-3-030-60460-8

This Springer imprint is published by the registered company Springer Nature Switzerland AG
The registered company address is: Gewerbestrasse 11, 6330 Cham, Switzerland

# Contents

# Brain-Computer Interface Research: A State-of-the-Art Summary 9

Christoph Guger, Michael Tangermann, and Brendan Z. Allison

**Abstract** Brain-computer interface (BCI) systems can provide communication and control without any physical movement. The BCI Research Awards are annual events to select the best BCI projects that year. Groups from around the world submit projects that are scored by a jury of international experts that selects twelve nominees and three winners. We also produce books like this one that review that year's nominees, awards ceremony, and winners. This introductory chapter briefly reviews BCIs and the 2019 awards process, including the jury, selection criteria, and nominees. We mention many chapters that might engage readers with different interests, including chapters with project descriptions or interviews with nominees. Many of the chapters here describe new approaches to BCIs that could be useful to patients and/or mainstream users. The final chapter of this book reviews the Awards Ceremony, announces the winners, and presents concluding comments.

**Keywords** Brain-computer interface · EEG · ECoG · BCI Research Awards · BCI Foundation · BCI Society

In the introduction to last year's book (Guger et al. 2020), we said that we were preparing for the Tenth Annual Brain-Computer Interface (BCI) Research Award ceremony as part of the 2020 BCI Meeting in Belgium. Since then, this conference has been postponed due to COVID. However, many entities have hosted online conferences, workshops, training sessions, and other events that show a strong ongoing interest in BCI research. With recent and upcoming online events from g.tec, NeurotechX, the organizers of the planned BCI Samara Conference, and other

C. Guger (✉)
g.tec medical engineering GmbH, Schiedlberg, Austria
e-mail: guger@gtec.at

M. Tangermann
Brain State Decoding Lab, Albert-Ludwigs-Universität Freiburg, Freiburg im Breisgau, Germany

B. Z. Allison
Cognitive Science Department, University of California at San Diego, La Jolla, USA

© The Author(s), under exclusive license to Springer Nature Switzerland AG 2021
C. Guger et al. (eds.), *Brain-Computer Interface Research*,
SpringerBriefs in Electrical and Computer Engineering,
https://doi.org/10.1007/978-3-030-60460-8_1

organizers, there are still many opportunities to become involved in BCI research or just learn more about the newest advances in our field.

We have also moved ahead with our ninth book, which is based on the BCI Research Award 2019. As with earlier books, we invited the authors of projects that were nominated for a BCI Research Award to contribute chapters describing what they did in their project, along with discussion and newer work from their group or other groups. Several authors in this book discuss next steps, future clinical directions, important challenges, and other issues to add breadth to their chapters.

# 1   What Is a BCI?

There is still no official, universally accepted definition of a brain-computer interfaces (BCI). Different articles have used slightly different definitions. However, journal papers and chapters that introduce or review BCIs have generally stated that BCIs are systems based on direct measures of brain activity that present real-time feedback to the end user. Real-time systems with advanced feedback are increasingly common, but the unique feature of BCIs is the reliance on brain signals that have not yet traveled elsewhere in the body (Wolpaw and Wolpaw 2012; Nam et al. 2018). The most widely cited review of BCIs states: "A BCI is a communication system in which messages or commands that an individual sends to the external world do not pass through the brain's normal output pathways of peripheral nerves and muscles. For example, in an EEG based BCI the messages are encoded in EEG activity. A BCI provides its user with an alternative method for acting on the world (Wolpaw et al. 2002)."

BCI systems do not need to rely exclusively on direct measures of brain activity. "Hybrid" BCIs might use BCIs along with other tools to convey information, including other BCIs, keyboards, mice, or systems based on speech, eye movement, or muscle activity. Hybrid BCIs began gaining attention in the literature about ten years ago (Lee et al. 2010; Müller-Putz et al. 2011; Leeb et al. 2011), and numerous more recent papers have presented or reviewed more advanced hybrid BCI systems (e.g., He et al. 2019; Rezazadeh Sereshkeh et al. 2019; Allison et al. 2020).

BCIs are also changing in terms of the people who can benefit from them. For many years, most BCI research sought to provide communication and/or control for persons with severe motor disabilities. Lou Gehrig's Disease, brainstem stroke, and other causes can leave people with little or no voluntary motor control. Since BCIs do not require voluntary motor control, they may be the only way for some people to interact with the outside world. Hence, keyboards, mice, and even some or all assistive technologies for disabled people may not be practical for them. However, more recent advances have shown that BCIs might be practical for different types of patients. As with prior books, the chapters in this book feature new ways to use BCIs to help broader patient groups.

BCIs for healthy users have been gaining attention as well. The past few years have seen high-profile announcements from Facebook and Elon Musk about large scale projects devoted to new BCI systems meant for healthy users. BCIs for healthy users are not new, and some applications meant for patients have also been validated with healthy users (Israel et al. 1980; Jung et al. 1997; Münßinger et al. 2010; Nijholt et al. 2019). However, most prior efforts have come from small research groups or companies with relatively limited resources. Hopefully, large-scale BCI efforts will push the field forward and foster new BCIs for healthy users and patients.

## 2 The Annual BCI Research Award

The Annual BCI Research Award is organized through the non-profit BCI Award Foundation. The Foundation was founded in 2017 in Austria and is chaired by Drs. Christoph Guger and Dean Krusienski. The BCI Award Foundation has Board Members to organize the Award. Editor Brendan Allison is also on the Board (Fig. 1).

Jury members may not submit projects. The award is open to any other research group, regardless of their location, equipment used, etc. The awards procedure this year followed a procedure like prior years:

- The BCI Award Foundation selects a Chairperson of the Jury from a top BCI research institute.
- The Chairperson selects a jury of international BCI experts to evaluate all projects submitted for the Award.
- The Award website[1] has instructions, scoring criteria, and the deadline for the Award.

Christoph Guger (AT)
President
g.tec medical engineering GmbH

Dean J. Krusienski (US)
Co-President
Virginia Commonwealth University

Tomasz M. Rutkowski (JP)
Treasurer
RIKEN AIP

Mikhail Lebedev (US)
Ambassador
Duke University Medical Center

Jing Jin (CN)
Ambassador
East China University of Science and Technology

Nuri Firat Ince (US)
Ambassador
University of Houston

**Fig. 1** The Board Members of the BCI Award Foundation

[1]https://www.bci-award.com/Home.

- The chairperson and BCI experts judge each submission.
- The jury chooses the first, second, and third place winners.
- The Award website announces the nominees.
- We ask the nominees to contribute a chapter to this annual book series, which may be a project summary and/or interview, and invite them to that year's Awards Ceremony within a major conference (such as an International BCI Meeting or Conference).
- Each Awards Ceremony is a major conference event.

The third-place prize was generously donated by the BCI Society. The BCI Society is a non-profit organization that organizes the BCI Meeting series (bcisociety.org). Authors CG and BA are members of the BCI Society and former Board Members. The other cash prizes were provided the Austrian company called g.tec medical engineering (author CG is the CEO), which manufactures equipment and software for BCIs and other applications.

The 2019 jury, shown in Fig. 2, included Dr. Ajiboye, who won the 2018 BCI Research Award, and Dr. Tangermann, last year's second place winner. Dr. Tangermann was also a nominee in 2018 and a juror in 2011. The 2019 jury also had a good mix of people with backgrounds in invasive and non-invasive BCIs who work in different areas active in BCIs. This prior experience and breadth are both important in juries, who need to evaluate a wide range of BCI projects each year.

The scoring criteria that the jury used to select the nominees and winners were the same as all previous BCI Research Awards:

The Jury of 2019

Michael Tangermann (DE)
Chair 2019
Albert-Ludwigs-University Freiburg
Brain State Decoding Lab

Abidemi Bolu Ajiboye (US)
Winner 2018
Case Western Reserve University Ohio

Rossella Spataro (IT)
University of Palermo

Michael Smith (US)
University of California, Berkeley

Keiichi Kitajo (JP)
RIKEN Center for Brain Science
National Institute for Physiological
Sciences

Selina Wriessnegger (AT)
Graz University of Technology
Institute of Neural Engineering

**Fig. 2** The jury for the 2019 BCI Research Award

- Does the project include a novel application of the BCI?
- Is there any new methodological approach used compared to earlier projects?
- Is there any new benefit for potential users of a BCI?
- Is there any improvement in terms of speed of the system (e.g. bit/min)?
- Is there any improvement in terms of accuracy of the system?
- Does the project include any results obtained from real patients or other potential users?
- Is the used approach working online/in real-time?
- Is there any improvement in terms of usability?
- Does the project include any novel hardware or software developments?

After the jury tallies the resulting scores, the nominees are posted online and invited to the Awards Ceremony. This ceremony has usually been part of the biggest BCI conference for that year. The BCI Society[2] coordinates BCI Meetings every even-numbered year, while the Technical University of Graz organizes a BCI Conference every odd-numbered year.[3]

This year's ceremony was part of the 8th Graz BCI Conference 2019. Like most years, the ceremony occurred in the evening to avoid conflict with daytime conference activities and provide a more relaxing atmosphere. The ceremony began with a short introduction to the BCI Awards and the selection procedure. Next, we asked one or more people whose project was nominated to join us onstage for a certificate. The ceremony concluded with announcing the first, second, and third place winners. The 1st place winner earns $3000 USD and the prestigious Gert Pfurtscheller bread knife trophy. The 2nd and 3rd place winners get $2000 USD and $1000 USD, respectively (Fig. 3).

## 3 The BCI Research Award Book Series

The first BCI Research Award was in 2010, and we've been producing a book along with the Award each year. Every year, we reviewed the main purpose of the awards and books. We want to recognize and encourage the top projects in BCI research worldwide. Each book contains chapters written by people who were nominated for that year's BCI Research Award. After the Awards Ceremony, we invite the nominees to write chapters about their work. Almost all chapters have reviewed the work nominated for the award. We provide the authors with several additional months after the ceremony to add new discoveries or results (from their group or other research groups), improved tables or figures, major challenges and possible solutions, future directions, and other commentary.

---

[2]Bcisociety.org.

[3]Tugraz.at/institutes/ine/home/.

**Fig. 3** The Chair of the Jury, Michael Tangermann, and jury member Selina Wriessnegger announce the projects nominated for the BCI Award at the Awards Ceremony at a BCI conference in Graz called the 8th Graz BCI Conference 2019

This year, for the first time, we had to leave the formatting entirely to Springer Publishing and their typesetters, with insufficient changes to proofs. The impact is obvious, and we hope the quality content shines through nonetheless.

Last year and this year, some chapters have been interviews, providing a different way to learn more about the nominee's project and related issues. For example, in chapter "Towards Brain-Machine Interface-Based Rehabilitation for Patients with Chronic Complete Paraplegia", we interviewed Dr. Solaiman Shokur, a Senior Researcher at Swiss Federal Institute of Technology in Lausanne (EPFL). Dr. Shokur discussed his team's research using EEG-based BCIs to help patients with spinal cord injury (SCI). Their BCI system included locomotion training and VR, and their results were the first to show that patients with certain types of neurological injuries could recover some brain function with this approach.

The introduction and discussion chapters are meant to be friendly and straight-forward. Readers who are new to BCIs, the BCI Research Awards, our book series, or the chapters in this year's book can learn more about all these topics. However, most of the chapters present more challenging material. Readers who are students or otherwise motivated to understand new terms and topics should be able to learn about BCI projects that interest them. Experts will also learn about some of the newest advances and the authors' perspectives. Interview chapters are often easier to read.

# 4  Projects Nominated for the BCI Award 2019

The twelve submissions with the highest scores were nominated for the BCI Research Award 2019. These nominees, affiliations, and project names were:

## 4.1  BCI-Based Neurofeedback Training for Quitting Smoking

Junjie Bu[1], Kymberly D. Young[2], Wei Hong[1], Ru Ma[1], Hongwen Song[5], Ying Wang[1], Wei Zhang[1], Michelle Hampson[3], Talma Hendler[4], Xiaochu Zhang[1,5].

[1]Hefei National Laboratory for Physical Sciences at the Microscale and School of Life Sciences, University of Science & Technology of China, Hefei, China.

[2]Department of Psychiatry, University of Pittsburgh School of Medicine, Pittsburgh, USA.

[3]Department of Radiology and Biomedical Imaging, Yale School of Medicine, New Haven, CT, USA.

[4]Functional Brain Center, Tel-Aviv University, Tel-Aviv, Israel.

[5]School of Humanities & Social Science, University of Science & Technology of China, Hefei, China.

## 4.2  Decoding Speech from Intracortical Multielectrode Arrays in Dorsal Motor Cortex

Sergey D. Stavisky[1], Francis R. Willett[1], Paymon Rezaii[1], Leigh R. Hochberg[2], Krishna V. Shenoy[1,3], Jaimie M. Henderson[1].

[1]Stanford University, USA.

[2]Brown University, Harvard Medical School, Massachusetts General Hospital, Providence VA Medical Center, USA.

[3]Howard Hughes Medical Institute, USA.

## 4.3  Neurofeedback of Scalp EEG Sensorimotor Rhythm Guides Hemispheric Activation of Sensorimotor Cortex

Masaaki Hayashi[1], Nobuaki Mizuguchi[2,3], Shohei Tsuchimoto[1,2], Shoko Kasuga[1,4,5], Junichi Ushiba[3,4].

[1]School of Fundamental Science and Technology, Graduate School of Keio University, Kanagawa, Japan.

[2]The Japan Society for the Promotion of Science, Tokyo, Japan.

[3]Department of Biosciences and informatics, Faculty of Science and Technology, Keio University, Kanagawa, Japan.
[4]Keio Institute of Pure and Applied Sciences, Kanagawa, Japan.
[5]Centre for Neuroscience Studies, Queen's University, Ontario, Canada.

## 4.4 Developing a Closed-Loop Brain-Computer Interface for Treatment of Neuropsychiatric Disorders Using Electrical Brain Stimulation

Yuxiao Yang[1], Omid G. Sani[1], Morgan B. Lee[2,3,4], Heather E. Dawes[2,3,4], Edward F. Chang[2,3,4], Maryam M. Shanechi[1,5].
[1]Ming Hsieh Department of Electrical Engineering, Viterbi School of Engineering, University of Southern California, USA.
[2]Department of Neurological Surgery, University of California, USA.
[3]Weill Institute for Neuroscience, University of California, San Francisco, USA.
[4]Kavli Institute for Fundamental Neuroscience, University of California, San Francisco, USA.
[5]Neuroscience Graduate Program, University of Southern California, USA.

## 4.5 Stentrode$^{TM}$ Neural Interface System: Minimally-Invasive Brain-Computer Interface Designed for Everyday Use

Peter Yoo[1], Nicholas Opie[1], Thomas Oxley[1], Stephen Ronayne[1], Gil Rind[1], Amos Meltzer[1].
[1]Synchron Inc., Australia.

## 4.6 Interfacing Hearing Implants with the Brain: Closing the Loop with Intracochlear Brain Recordings

Ben Somers[1], Damien Lesenfants[1], Jonas Vanthornhout[1], Lien Decruy[1], Eline Verschueren[1], Tom Francart[1].
[1]KU Leuven—University of Leuven, Department of Neurosciences, ExpORL, Leuven, Belgium.

## 4.7 A Brain–Spine Interface Alleviating Gait Deficits in a Primate Model of Parkinson's Disease

Tomislav Milekovic[1,2], Flavio Raschellà[3], Matthew G. Perich[2], Shiqi Sun[1,4], Eduardo Martin Moraud[6,7], Giuseppe Schiavone[5], Yang Jianzhong[8,9], Andrea Galvez[2], Christopher Hitz[1], Alessio Salomon[1], David Borton[1,10], Jean Laurens[1,11], Isabelle Vollenweider[1], Simon Borgognon[1], Jean-Baptiste Mignardot[1], Wai Kin D Ko[8,9], Cheng YunLong[8,9], Li Hao[8,9], Peng Hao[8,9], Qin Li[8,9], Marco Capogrosso[12], Tim Denison[13], Stéphanie P. Lacour[5], Silvestro Micera[3,14], Chuan Qin[9], Jocelyne Bloch[6,7], Erwan Bezard[8,10,15,16], Grégoire Courtine[1,6,7].

[1]Center for Neuroprosthetics and Brain Mind Institute, School of Life Sciences, EPFL, Switzerland.

[2]Department of Fundamental Neuroscience, Faculty of Medicine, University of Geneva, Switzerland.

[3]Center for Neuroprosthetics and Institute of Bioengineering, School of Engineering, EPFL, Switzerland.

[4]Beijing Engineering Research Center for Intelligent Rehabilitation, College of Engineering, Peking University, People's Republic of China.

[5]Center for Neuroprosthetics, Institute of Microengineering and Institute of Bioengineering, School of Engineering, EPFL, Switzerland.

[6]Department of Clinical Neuroscience, Lausanne University Hospital (CHUV) and University of Lausanne, Switzerland.

[7]Department of Neurosurgery, CHUV, Switzerland.

[8]Motac Neuroscience, United Kingdom.

[9]Institute of Laboratory Animal Sciences, China Academy of Medical Sciences, People's Republic of China.

[10]Carney Institute for Brain Science, School of Engineering, Brown University, USA.

[11]Department of Neuroscience, Baylor College of Medicine, USA.

[12]Translational Neuroscience platform, University of Fribourg, Switzerland.

[13]Oxford University, United Kingdom.

[14]The BioRobotics Institute, Scuola Superiore Sant'Anna, Italy.

[15]Université de Bordeaux, Institut des Maladies Neurodégénératives (IMN), France.

[16]CNRS, IMN, France.

## 4.8   Post-stroke Rehabilitation Training with a Motor-Imagery-Based Brain-Computer Interface (BCI)-Controlled Hand Exoskeleton: A Randomized Controlled Multicenter Trial

Frolov Alexander[1,2], Biryukova Elena[1,2], Bobrov Pavel[1,2], Bobrov Dmirty[1], Lekin Alexander[1], Mokienko Olesya[3], Lyukmanov Roman[3], Kotov Sergey[4], Kondur Anna[4], Ivanova Galina[1], Bushkova Yulia[1].
[1]Pirogov Russian National Research Medical University, Russia.
[2]Institute of Higher Nervous Activity, Russia.
[3]Research Centre of Neurology, Russia.
[4]Vladimirsky Moscow Regional Research Clinical Institute, Russia.

## 4.9   The Walk Again Neurorehabilitation Protocol: A BMI-Based Clinical Application to Induce Partial Neurological Recovery in Spinal Cord Injury Patients

Solaiman Shokur[1], Debora S. F. Campos[1], A. R. C. Donati[1], Eduardo J. L. Alho[1], Mikhail Lebedev[1], Miguel Nicolelis[1].
[1]Neurorehabilitation laboratory AASDAP.

## 4.10   Hearables: In-Ear Multimodal Brain Computer Interfacing

Metin Yarici[1], Harry J. Davies[1], Takashi Nakamura[1], Ian Williams[1], Danilo P. Mandic[1].
[1]Imperial College London, UK.

## 4.11   Power Modulations of ECoG Alpha/Beta and Gamma Bands Correlate with Time Derivative of Force During Sustained Hand Grasp

Tianxiao Jiang[1], Priscella Asman[1], Giuseppe Pellizzer[2], Sudhakar Tummala[3], Sujit Prabhu[3], Nuri F. Ince[1].
[1]University of Houston, USA.
[2]University of Minnesota, USA.
[3]MD Anderson Cancer Center, USA.

### 4.12 Next-Generation Microscale Wireless Implant System for High-Density, Multi-areal Closed-Loop Brain Computer Interfaces

Farah L. Laiwalla[1], Vincent W. Leung[2], Jihun Lee[1], Patrick Mercier[2], Peter Asbeck[2], Ramesh Rao[2], Lawrence Larson[1], Arto Nurmikko[1].
[1]Brown University, USA.
[2]University of California San Diego, USA.

## 5 Summary

The subsequent chapters in this book present interviews and research directions that may interest a myriad of different readers. Like chapters from preceding books, this year's chapters include both invasive and non-invasive BCIs, with different system components, user interaction paradigms, signal processing methods, and goals. Many chapters present new approaches to help different patient groups.

For example, neurofeedback was prominent in several projects nominated in 2019. Chapters "BCI-Based Neurofeedback Training for Quitting Smoking" and "Training with BCI-Based Neurofeedback for Quitting Smoking" present a novel type of BCI for quitting smoking, with one chapter focused on their project and another chapter devoted to an interview. Their project used well-established EEG neurofeedback principles combined with advanced BCI techniques that was effective in a double-blind trial. Chapters "Developing a Closed-Loop Brain-Computer Interface for Treatment of Neuropsychiatric Disorders Using Electrical Brain Stimulation" and "Closed-Loop BCI for the Treatment of Neuropsychiatric Disorders" describe a different approach to use neurofeedback for patients with neuropsychiatric disorders. The project in chapter "Neurofeedback of Scalp Bi-Hemispheric EEG Sensorimotor Rhythm Guides Hemispheric Activation of Sensorimotor Cortex in the Targeted Hemisphere" addressed neurofeedback for sensorimotor control. Different chapters present BCI systems to help patients recover from stroke, produce or understand speech, or use tactile feedback to support grasping and other activities. Like many directions presented in different chapters, their work is not yet ready for widespread clinical application, but could ignite new ideas and follow-up efforts that could help many people. The last chapter of this book presents the winners of the 2019 BCI Research Award and some discussion.

## References

B.Z. Allison, A. Kübler, J. Jin, 30+ years of P300 brain-computer interfaces. Psychophysiology **57**, e13569–e13569 (2020)

C. Guger, K. Miller, B.Z. Allison, Introduction, in *Brain–Computer Interface Research. Springer-Briefs in Electrical and Computer Engineering*, ed. by C. Guger, B.Z. Allison, K. Miller (Springer, Cham, 2020). https://doi.org/10.1007/978-3-030-49583-1_1

S. He, Y. Zhou, T. Yu, R. Zhang, Q. Huang, L. Chuai, Y. Li, EEG-and EOG-Based Asynchronous Hybrid BCI: A System Integrating a Speller, a Web Browser, an E-Mail Client, and a File Explorer. IEEE Trans. Neural Syst. Rehabil. Eng. **28**(2), 519–530 (2019)

J. Israel, C.D. Wickens, G. Chesney, E. Donchin, The event related brain potential as a selective index of display monitoring load. Hum. Factors **22**, 280–294 (1980)

T.P. Jung, S. Makeig, M. Stensmo, T.J. Sejnowski, Estimating alertness from the EEG power spectrum. IEEE Trans. Biomed. Eng. **44**(1), 60–69 (1997)

E.C. Lee, J.C. Woo, J.H. Kim, M. Whang, K.R. Park, A brain–computer interface method combined with eye tracking for 3D interaction. J. Neurosci. Methods **190**(2), 289–298 (2010)

R. Leeb, H. Sagha, R. Chavarriaga, J. del R Millán, A hybrid brain–computer interface based on the fusion of electroencephalographic and electromyographic activities. J. Neural Eng. **8**(2), 025011 (2011)

G.R. Müller-Putz, C. Breitwieser, F. Cincotti, R. Leeb, M. Schreuder, F. Leotta, M. Tavella, L. Bianchi, A. Kreilinger, A. Ramsay, M. Rohm, Tools for brain-computer interaction: A general concept for a hybrid BCI. Front. Neuroinf. **5**, 30 (2011)

J.I. Münßinger, S. Halder, S.C. Kleih, A. Furdea, V. Raco, A. Hösle, A. Kubler, Brain painting: first evaluation of a new brain–computer interface application with ALS-patients and healthy volunteers. Front. Neurosci. **4**, 182 (2010)

C.S. Nam, A. Nijholt, F. Lotte, Introduction: evolution of brain-computer interfaces, in *Brain-Computer Interfaces Handbook: Technological and Theoretical Advances* (CRC Press is an imprint of the Taylor & Francis Group, Boca Raton, 2018)

A. Nijholt (ed.), *Brain Art: Brain-Computer Interfaces for Artistic Expression* (Springer, Cham, 2019)

A. Rezazadeh Sereshkeh, R. Yousefi, A.T. Wong, F. Rudzicz, T. Chau, Development of a ternary hybrid fNIRS-EEG brain–computer interface based on imagined speech. Brain-Comp. Interfaces **6**(4), 128–140 (2019)

J.R. Wolpaw, N. Birbaumer, D.J. McFarland, G. Pfurtscheller, T.M. Vaughan, Brain–computer interfaces for communication and control. Clin. Neurophysiol. **113**(6), 767–791 (2002)

J.R. Wolpaw, E.W. Wolpaw, Brain-computer interfaces: something new under the sun, in *Brain-Computer Interfaces: Principles and Practice*, vol. 14 (Oxford University Press, Oxford, 2012)

# BCI-Based Neurofeedback Training for Quitting Smoking

**Junjie Bu and Xiaochu Zhang**

**Abstract** Neurofeedback is a psychophysiological protocol in which online feed-back of brain activation is provided to the participant for self-regulation. As a progenitor of brain–computer interfaces (BCIs), neurofeedback has provided a novel way to improve brain function and investigate neuroplasticity. Previous EEG-based neurofeedback protocols have been employed in drug addiction treatment for more than four decades. However, the efficacy of these traditional EEG neurofeedback approaches in the treatment of addiction remains dubious. Here we developed a novel cognition-guided neurofeedback protocol and evaluated its therapeutic efficacy on nicotine addiction. We trained a personalized multivariate pattern analysis (MVPA) classifier to identify an EEG activity pattern associated with drug cue reactivity using the specific cognitive task (drug cue reactivity task) before neurofeedback, and subsequently trained participants to de-activate this pattern during neurofeedback (hereby termed 'cognition-guided neurofeedback'). In a double-blind, placebo-controlled, randomized clinical trial, 60 nicotine-dependent participants were assigned to receive two neurofeedback training sessions (about 1 h/session) either from their own brain ($N = 30$, real-feedback group) or from the brain activity pattern of a matched participant ($N = 30$, yoked-feedback group). In the real-feedback group, participants successfully de-activated EEG activity patterns of smoking cue reactivity. The real-feedback group showed significant decrease in cigarette craving and craving-related P300 amplitudes compared with the yoked-feedback group. The rates of cigarettes smoked per day at 1-week, 1-month and 4-month follow-up decreased 30.6, 38.2, and 27.4% relative to baseline in the real-feedback group, compared to decreases of 14.0, 13.7, and 5.9% in the yoked-feedback group. The neurofeedback effects on craving change and smoking amount at the 4-month follow-up were further predicted by neural markers at pre-neurofeedback. This novel neurofeedback training approach produced significant short-term and long-term effects on cigarette craving

J. Bu (✉)
School of Biomedical Engineering, Research and Engineering Center of Biomedical Materials, Anhui Medical University, Hefei, China
e-mail: bujunjie@ahmu.edu.cn

J. Bu · X. Zhang
School of Life Sciences, University of Science & Technology of China, Hefei, China

© The Author(s), under exclusive license to Springer Nature Switzerland AG 2021
C. Guger et al. (eds.), *Brain-Computer Interface Research*,
SpringerBriefs in Electrical and Computer Engineering,
https://doi.org/10.1007/978-3-030-60460-8_2

and smoking behavior, suggesting the neurofeedback protocol described herein is a promising brain-based tool for treating addiction.

**Keywords** Brain-computer interface · Cognition-guided neurofeedback · Nicotine addiction · Smoking cue reactivity

# 1  Introduction

Nicotine addiction is the leading preventable cause of disease and death worldwide. With approximately 75% of patients with nicotine dependence not responding fully to the Gold Standard Programme (a comprehensive intervention consisting of manual-based teaching sessions together with nicotine replacement therapy) for smoking cessation interventions (Rasmussen et al. 2017), high relapse rates during long-term follow-up periods remain a core feature of nicotine addiction. Therefore, there is an urgent need to develop novel therapeutic approaches for nicotine addiction.

Neurofeedback, a psychophysiological procedure that helps participants self-regulate their brain activity, has been of growing interest among basic and clinical neuroscientists (Sitaram et al. 2017). Clinically, neurofeedback has been employed in many psychiatric disorders, including Attention Deficit Hyperactivity Disorder (Arns et al. 2009), depression (Young et al. 2017), anxiety (Scheinost et al. 2013) and drug addiction (Sokhadze et al. 2008). Further, recent fMRI-based neurofeedback studies indicate preliminary efficacy in reducing cigarette craving in smokers (Li et al. 2013; Hartwell et al. 2016). However, the feasibility of turning fMRI neurofeedback into a widely available clinical intervention is questionable. In contrast, EEG is a relatively inexpensive and portable brain imaging technique that can be easily implemented at any location and has more potential for wide-spread clinical implementation than fMRI neurofeedback. Previous EEG-based neurofeedback protocols, including alpha training, alpha/theta training, and SMR (sensorimotor rhythm)/beta training, have been employed in drug addiction treatment for more than four decades (Sokhadze et al. 2008). Using these training protocols, drug-dependent patients received the power of a single and fixed EEG frequency and self-regulated that signal (Schmidt et al. 2017). Most of these studies focused on facilitating relaxation and reducing anxiety. However, the efficacy on drug addiction has only been classified as "probably efficacious" in reports from the Association for Applied Psychophysiology and Biofeedback and the International Society for Neurofeedback and Research (Sokhadze et al. 2008; Schmidt et al. 2017). Additionally, recent studies question the clinical efficacy of previous EEG neurofeedback protocols (Thibault and Raz 2016; Fovet et al. 2017; Schabus et al. 2017). Hence, a new direction for EEG neurofeedback in treating drug addiction is warranted.

The efficacy of these traditional EEG neurofeedback approaches in the treatment of addiction remains dubious, in part, because addiction process involves many complex cognitive models (e.g. the cue reactivity model (Chiamulera 2005) and negative reinforcement model Koob [2013]), but previous neurofeedback studies mainly targeted arousal and/or anxiety. Instead, drug cue reactivity can evoke the impulse for drug-seeking behavior in addiction (Weiss et al. 2001). Previous studies from our group and others indicate that smoking cue reactivity is a central characteristic of nicotine addiction (Zhang et al. 2009; Engelmann et al. 2012) and can predict relapse vulnerability (Janes et al. 2010); thus, there is evidential support for a causal relationship between cue reactivity and relapse (Parvaz et al. 2011). Therefore, reducing brain reactivity to smoking cues has the potential to improve smoking cessation outcomes.

Recent EEG studies have reported that smoking cue reactivity is a complex brain activity pattern that involves multiple EEG features, including both time domain (e.g., P300, slow positive wave) and frequency domain (e.g., alpha oscillation) features (Cui et al. 2013; Littel and Franken 2007; Littel et al. 2012). Typically, multivariate pattern analysis (MVPA) can enhance sensitivity of detecting a particular brain activity pattern by using multifeature combinations for input to multivariate patterns (Haynes and Rees 2006). A number of neurofeedback studies combined with MVPA have been impressively successful at improving attention and perceptual learning after only a few sessions (deBettencourt et al. 2015; Shibata et al. 2011), whereas some traditional EEG-based neurofeedback studies require dozens of sessions for any effects to be detected (Sokhadze et al. 2008).

In the current study, we evaluated a novel EEG neurofeedback paradigm (cognition-guided neurofeedback) effects on nicotine addiction by double-blind, randomized, placebo-controlled design.

## 2 Cognition-Guided Neurofeedback

The cognition-guided neurofeedback training paradigm consisted of two parts (Fig. 1a, b). First, we trained a personalized classifier to distinguish the EEG activity patterns corresponding to smoking and neutral cue reactivity using the smoking cue reactivity task. Raw signals were pre-processed offline based on the EEG signals acquired during the smoking cue reactivity task. Afterwards, time domain (EEG potential) and time-frequency domain (EEG power) features were calculated by contrasting smoking with neutral cues using a permutation test. Time-frequency domain features were calculated by wavelet analysis. Under the threshold ($p < 0.05$) of the permutation test, the EEG potential and EEG power features surviving this threshold were separately formed into temporal-spatial clusters by grouping them at adjacent time points and electrodes using a cluster-based statistic (Groppe et al. 2011). Once the temporal-spatial clusters were identified, the EEG features for constructing the pattern classifier were extracted from these clusters. The mean values (potential and power) from each temporal-spatial cluster were calculated and combined into a linear support vector machine (SVM) classifier. The SVM classifier was selected since it often outperforms other classifiers for neurofeedback (Lotte et al. 2018).

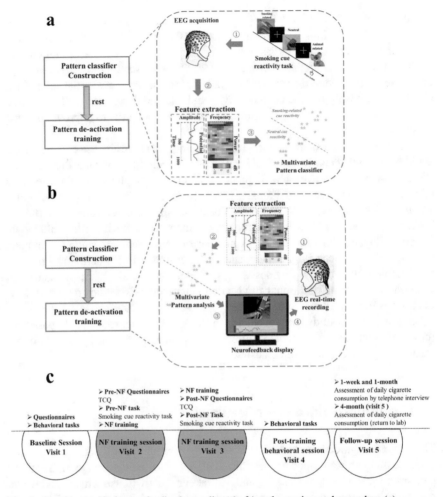

**Fig. 1** Cognition-guided neurofeedback paradigm (**a**, **b**) and experimental procedure (**c**)

Next, during neurofeedback training, participants were asked to repeatedly and continuously de-activate their real-time EEG activity patterns of smoking cue reactivity calculated using a previously constructed classifier. For each real-time raw EEG signal lasting 1 s, the pre-processing was used by the same algorithm as in the previous EEG pre-processing. The time domain and time-frequency domain features were then extracted from the previous temporal-spatial clusters by calculating the mean values (EEG potential and EEG power), and then input into the personalized classifier. The classifier estimated the probabilistic score in real time reflecting the extent to which the brain activity pattern matched the pattern for reactivity to the smoking cue. We determined the probabilistic scores (from 0 to 1) for the classifier. As a result, when a participant successfully deactivated the smoking cue pattern, the probabilistic score decreased. That is, when the current activity patterns were more similar to neutral cue activity patterns, the score decreased, and when the current activity patterns were more similar to the smoking cue activity patterns, the score increased.

# 3  Adaptive Closed-Loop Design

To improve participants' vigilance, and help them better self-monitor and evaluate their brain state during neurofeedback, we used an adaptive closed-loop method in which the neurofeedback stimulus and decoded brain state influenced each other in real-time (deBettencourt et al. 2015). Different craving level pictures evoked different degrees of smoking cue reactivity for smokers (Carter and Tiffany 1999). In the current study, an approach of continually updating sensory stimuli (e.g., different craving level pictures) based on changing brain states (e.g., different degrees of smoking cue reactivity pattern) constituted a "closed-loop" design. The logic of this closed-loop design is that, when a participant was unsuccessful in "deactivating" the smoking cue pattern (i.e., the probabilistic score increased), a picture with a higher craving level was displayed to amplify and externalize the consequences of the participant's smoking cue related brain activity pattern (deBettencourt et al. 2015; deBettencourt and Norman 2016). This made unsuccessful deactivation more salient and increased the self-monitoring demand of the task. In other words, we amplified the consequences of their cue pattern, rewarding successful deactivation by reducing difficulty with a lower craving level picture and punishing unsuccessful deactivation by increasing difficulty with a higher craving level picture.

The probabilistic score was presented at the bottom half of the screen and transferred into a picture presented at the top half of the screen at the same time (Fig. 1b). The association between the probabilistic score and the transferred picture was controlled by a linear positive correlation function. To reduce fluctuations due to noise in the EEG signal, the probabilistic score value of each trial was calculated using a moving average of the current and two preceding values.

After neurofeedback practice, participants were required to identify ten cognitive strategies that may be effective at de-activating the neurofeedback signal, but it was emphasized that they should adjust their strategies to find a method that works best for them during neurofeedback (Instruction: "Make the feedback curve move down and the picture induce less craving"). Each neurofeedback training session consisted of 8 cycles, with 40 trials per cycle. Each trial was updated every 2 s including 1 s acquisition and 1 s computing, with a 1 min rest between cycles. At the end of each cycle, the self-regulation performance during the previous cycle was presented. After each cycle, participants rated their perceived control over the neurofeedback signal. The final cumulative performance was translated into an additional money reward. Both groups received the same money after neurofeedback.

## 4   Experimental Procedure and Participants

The study was a double blind, randomized, placebo-controlled design. The experimental procedure consisted of four stages (Fig. 1c): (1) baseline session (Visit 1); (2) two neurofeedback training sessions (Visit 2 and Visit 3); (3) post-training behavioral session (Visit 4); and (4) follow-up session (Visit 5). Participants were required to be abstinent from smoking cigarettes for two hours prior to every visit, which ensured participants had some craving and responsiveness to the cues without the potential confound of a ceiling effect from prolonged abstinence.

Sixty participants who met the following criteria participated in the experiment: smoking 10 or more cigarettes per day for 2 years or more, right-handed, between 18 and 40 years of age, normal or corrected to normal vision, and in good mental and physical health assessed by the Mini-International Neuropsychiatric Interview (Sheehan et al. 1998).

Participants were randomly assigned to the real-feedback group ($n = 30$) or the yoked-feedback group ($n = 30$). The real-feedback group regulated their own online brain patterns. The yoked-feedback group regulated the brain activity pattern of a matched participant in the real-feedback group (deBettencourt et al. 2015).

## 5   Cognition-Guided Neurofeedback Effects on Nicotine Addiction

Figure 2 indicates that the real-feedback group successfully de-activated smoking cue reactivity patterns after two neurofeedback visits. A linear regression analysis revealed that there was a strong and significant negative correlation between training cycle and the mean probabilistic score across participants in the real-feedback group ($r = -0.155, p = 0.001$) (Fig. 2a). However, this finding was not observed in the yoked-feedback group ($r = 0.015, p = 0.77$) (Fig. 2b) and the correlation was significantly different from the real-feedback group ($z = -2.47, p = 0.013$). A two-way mixed-design ANOVA using group (real-feedback, yoked-feedback) as a between-subjects factor and cycle (first cycle of neurofeedback visit 1, last cycle of neurofeedback visit 2) as a within-subjects factor on the probabilistic score revealed a significant group-by-cycle interaction ($F(1, 51) = 4.04, p = 0.04, d = 0.56$).

A two-way mixed-design ANOVA using group (real-feedback, yoked-feedback) as a between-subjects factor and time (pre-neurofeedback, post-neurofeedback) as a within-subjects factor on the cigarette craving score revealed a significant group by time interaction ($F(1, 51) = 4.69, p = 0.03, d = 0.61$) (Fig. 3a). Furthermore, participants in the real-feedback group with lower levels of average de-activated neurofeedback performance exhibited greater decreases in craving scores ($r = -0.40, p = 0.03$), which was consistent with our hypothesized mechanism of action for this intervention (Fig. 3b). This correlation was not significant in the yoked-feedback group ($r = -0.12, p = 0.53$). In addition, we

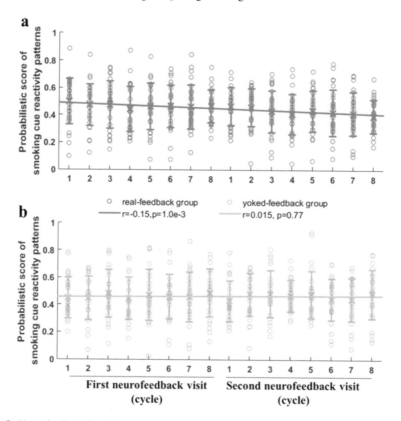

**Fig. 2** Neurofeedback learning

compared the craving-related P300 component evoked by smoking-related cues. A two-way mixed-design ANOVA using group (real-feedback, yoked-feedback) as a between-subjects factor and time (pre-neurofeedback, post-neurofeedback) as a within-subjects factor on average amplitude at selected group peak time window (350 ms ~ 450 ms) revealed a significant group-by-time interaction effect ($F(1, 51) = 5.13$, $p = 0.028$, $d = 0.64$). These findings indicated neurofeedback produced short-term effects on cigarette craving.

A two-way mixed-design ANOVA using group (real-feedback, yoked-feedback) as a between-subjects factor and time (pre-neurofeedback, 1-week follow-up, 1-month follow-up, 4-month follow-up visit) as a within-subjects factor on daily cigarette consumption revealed a significant group by time interaction ($F(3, 126) = 3.68$, $p = 0.01$, $d = 0.59$) (Fig. 3c). After two neurofeedback training visits, the real-feedback group showed significantly decreased cigarette consumption per day compared to the yoked-feedback group at the 1-week follow-up ($t(48) = -2.53$, $p = 0.01$, $d = 0.72$), 1-month follow-up ($t(47) = -2.98$, $p < 0.005$, $d = 0.86$), and 4-month follow-up ($t(42) = -2.21$, $p = 0.03$, $d = 0.67$). In addition,

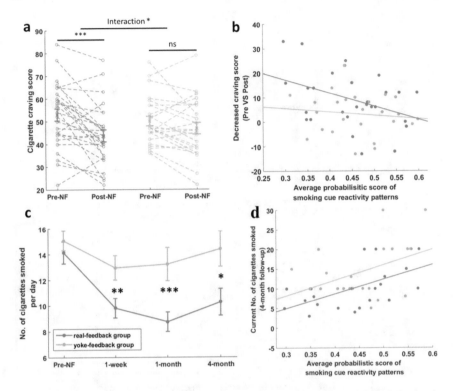

**Fig. 3** Neurofeedback effects on short-term cigarette craving and long-term smoking behavior. *p < 0.05; **p < 0.01; ***p < 0.005; ns: not significant. NF: neurofeedback

the real-feedback group showed a significant correlation between the average de-activated neurofeedback performance and the current cigarette amount at 4-month follow-up ($r = 0.58$, $p = 0.004$) (Fig. 3d). These findings indicated neurofeedback produced long-term effects on smoking behavior.

Figure 4a shows that the classification accuracy of the pre-neurofeedback classifier significantly predicted decreased craving scores in the real-feedback group ($r = 0.40$, $p = 0.03$), while the same prediction was not significant in the yoked-feedback group ($r = 0.13$, $p = .54$). Moreover, the correlation analysis revealed that the degree of de-activation during the first cycle of the first neurofeedback successfully predicted the number of cigarettes smoked per day at the 4-month follow-up ($r = 0.45$, $p = 0.03$, Fig. 4b) in the real-feedback group, but not in the yoked-feedback group ($r = 0.16$, $p = 0.45$). These findings indicated short- and long-term effects were predicted by the classification accuracy at pre-neurofeedback and neurofeedback performance during the first training cycle, respectively.

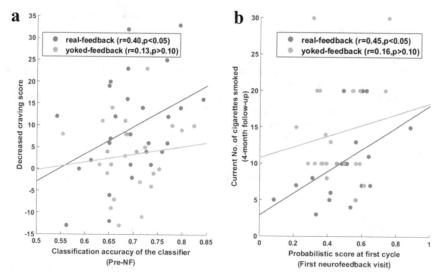

**Fig. 4** Individual-level prediction of short-term (**a**) and long-term (**b**) effects. NF: neurofeedback

## 6 Conclusion

In conclusion, we developed and tested a novel cognition-guided neurofeedback protocol to de-activate EEG activity patterns of smoking cue reactivity, which produced short- and long-term effects on cigarette craving and smoking behavior. In particular, the rate of smoking amount decreased as much as 38.2% during the 4-month follow-up period after only 2 h of neurofeedback training. These results suggest this novel neurofeedback intervention is a promising treatment for addiction, with potential to be a low-cost and high-portability brain-based treatment for addiction. This approach therefore merits further testing.

## References

M. Arns et al., Efficacy of neurofeedback treatment in ADHD: the effects on inattention, impulsivity and hyperactivity: a meta-analysis. Clin. EEG Neurosci. **40**(3), 180–189 (2009)

B.L. Carter, S.T. Tiffany, Meta-analysis of cue-reactivity in addiction research. Addiction **94**(3), 327–340 (1999)

C. Chiamulera, Cue reactivity in nicotine and tobacco dependence: a "multiple-action" model of nicotine as a primary reinforcement and as an enhancer of the effects of smoking-associated stimuli. Brain Res. Rev. **48**(1), 74–97 (2005)

Y. Cui et al., Alpha oscillations in response to affective and cigarette-related stimuli in smokers. Nicotine Tob. Res. **15**(5), 917–924 (2013)

M.T. deBettencourt, K.A. Norman, Neuroscience: Incepting associations. Curr. Biol. **26**(14), R673–R675 (2016)

M.T. deBettencourt et al., Closed-loop training of attention with real-time brain imaging. Nat. Neurosci. **18**(3), 470–475 (2015)

J.M. Engelmann et al., Neural substrates of smoking cue reactivity: a meta-analysis of fMRI studies. Neuroimage **60**(1), 252–262 (2012)

T. Fovet et al., On assessing neurofeedback effects: should double-blind replace neurophysiological mechanisms? Brain **140**(10), e63 (2017)

D.M. Groppe, T.P. Urbach, M. Kutas, Mass univariate analysis of event-related brain potentials/fields I: a critical tutorial review. Psychophysiology **48**(12), 1711–1725 (2011)

K.J. Hartwell et al., Individualized real-time fMRI neurofeedback to attenuate craving in nicotine-dependent smokers. J. Psychiatry Neurosci. **41**(1), 48–55 (2016)

J.D. Haynes, G. Rees, Decoding mental states from brain activity in humans. Nat. Rev. Neurosci. **7**(7), 523–534 (2006)

A.C. Janes et al., Brain reactivity to smoking cues prior to smoking cessation predicts ability to maintain tobacco abstinence. Biol. Psychiatry **67**(8), 722–729 (2010)

G.F. Koob, Negative reinforcement in drug addiction: the darkness within. Curr. Opin. Neurobiol. **23**(4), 559–563 (2013)

X. Li et al., Volitional reduction of anterior cingulate cortex activity produces decreased cue craving in smoking cessation: a preliminary real-time fMRI study. Addict. Biol. **18**(4), 739–748 (2013)

M. Littel, I.H. Franken, The effects of prolonged abstinence on the processing of smoking cues: an ERP study among smokers, ex-smokers and never-smokers. J. Psychopharmacol **21**(8), 873–882 (2007)

M. Littel et al., Electrophysiological indices of biased cognitive processing of substance-related cues: a meta-analysis. Neurosci. Biobehav. Rev. **36**(8), 1803–1816 (2012)

F. Lotte et al., A review of classification algorithms for EEG-based brain-computer interfaces: a 10 year update. J. Neural Eng. **15**(3), 031005 (2018)

M.A. Parvaz et al., Neuroimaging for drug addiction and related behaviors. Rev. Neurosci. **22**(6), 609–624 (2011)

M. Rasmussen, E. Fernandez, H. Tonnesen, Effectiveness of the Gold Standard Programme compared with other smoking cessation interventions in Denmark: a cohort study. BMJ Open **7**(2), e013553 (2017)

M. Schabus, et al., Better than sham? A double-blind placebo-controlled neurofeedback study in primary insomnia. Brain (2017)

D. Scheinost et al., Orbitofrontal cortex neurofeedback produces lasting changes in contamination anxiety and resting-state connectivity. Transl. Psychiatry **3**, e250 (2013)

J. Schmidt, C. Kärgel, M. Opwis, Neurofeedback in substance use and overeating: current applications and future directions. Curr. Addict. Rep. **4**(2), 116–131 (2017)

D.V. Sheehan, et al., The Mini-International Neuropsychiatric Interview (M.I.N.I.): the development and validation of a structured diagnostic psychiatric interview for DSM-IV and ICD-10. J. Clin. Psychiatry **59**(Suppl 20), 22–33;quiz 34–57 (1998)

K. Shibata et al., Perceptual learning incepted by decoded fMRI neurofeedback without stimulus presentation. Science **334**(6061), 1413–1415 (2011)

R. Sitaram et al., Closed-loop brain training: the science of neurofeedback. Nat. Rev. Neurosci. **18**(2), 86–100 (2017)

T.M. Sokhadze, R.L. Cannon, D.L. Trudeau, EEG biofeedback as a treatment for substance use disorders: Review, rating of efficacy, and recommendations for further research. Appl. Psychophysiol. Biofeedback **33**(1), 1–28 (2008)

R.T. Thibault, A. Raz, When can neurofeedback join the clinical armamentarium? Lancet Psychiatry **3**(6), 497–498 (2016)

F. Weiss et al., Compulsive drug-seeking behavior and relapse - Neuroadaptation, stress, and conditioning factors. Biol. Basis Cocaine Addict. **937**, 1–26 (2001)

K.D. Young et al., Randomized Clinical Trial of Real-Time fMRI Amygdala Neurofeedback for Major Depressive Disorder: Effects on Symptoms and Autobiographical Memory Recall. Am. J. Psychiatry **174**(8), 748–755 (2017)

X. Zhang et al., Masked smoking-related images modulate brain activity in smokers. Hum. Brain Mapp. **30**(3), 896–907 (2009)

# Neurofeedback of Scalp Bi-Hemispheric EEG Sensorimotor Rhythm Guides Hemispheric Activation of Sensorimotor Cortex in the Targeted Hemisphere

**Masaaki Hayashi, Nobuaki Mizuguchi, Shohei Tsuchimoto, and Junichi Ushiba**

**Abstract** Oscillatory electroencephalographic (EEG) activity is associated with excitability of cortical regions. Visual feedback of EEG-oscillations may promote increased excitability in targeted cortical regions, but is not truly guaranteed due to its limited spatial specificity and signal interaction among interhemispheric brain regions. Guiding spatially specific sensorimotor cortical activation is important for facilitating neural rehabilitation processes. Here, we developed a spatially bivariate EEG-based neurofeedback approach that monitors bi-hemispheric sensorimotor activities during unilateral upper-limb motor imagery (MI), and tested whether users could volitionally lateralize sensorimotor activity to the contralateral or ipsilateral hemisphere using right shoulder MI-associated neurofeedback. Then, hand MI-associated BCI-neurofeedback was tested as a negative control via the same procedure. Lateralized EEG activity was compared between shoulder and hand MIs to see how differences in intrinsic corticomuscular projection patterns might influence activity lateralization. In right shoulder MI, ipsilaterally and contralaterally dominant sensorimotor activation was guided via EEG-based neurofeedback. Conversely, in right hand MI, only contralaterally (but not ipsilaterally) dominant sensorimotor activation was guided. These results are compatible with neuroanatomy;

This article is a summary of the paper in Neuroimage (https://doi.org/10.1016/j.neuroimage.2020. 117298) and is licensed by Elsevier.

M. Hayashi · S. Tsuchimoto
School of Fundamental Science and Technology, Graduate School of Keio University, Kanagawa, Japan

N. Mizuguchi · S. Tsuchimoto
The Japan Society for the Promotion of Science, Tokyo, Japan

N. Mizuguchi · J. Ushiba (✉)
Department of Biosciences and Informatics, Faculty of Science and Technology, Keio University, Kanagawa, Japan
e-mail: ushiba@brain.bio.keio.ac.jp

J. Ushiba
Keio Institute of Pure and Applied Sciences, Kanagawa, Japan

shoulder muscles are innervated bihemispherically, whereas hand muscles are mostly innervated contralaterally.

**Keywords** Brain-computer interface · Motor imagery · Laterality · Sensorimotor cortical activity · Neural plasticity

# 1 Introduction

Oscillatory brain activity is associated with the excitability of a cortical region. Changes in frequency, amplitude, and phase of ongoing oscillation cycles visualized in electro- or magnetoencephalograms (EEG/MEG) over the human sensorimotor cortex (SM1) are associated with modulated responses in SM1. In particular, the excitability of SM1 and the connected spinal motoneuron pool is significantly higher when the amplitude of sensorimotor rhythms (SMR) in the alpha (8–13 Hz) and beta (14–30 Hz) bands is lower (Neuper and Pfurtscheller 2001; Neuper et al. 2006; Pfurtscheller et al. 2006). This concept has inspired neurofeedback interventions via a brain-computer interface (BCI) whereby, for example, post-stroke hemiplegic patients learn to volitionally desynchronize/synchronize SMR signals in the ipsilesional hemisphere through visual or sensory feedbacks, with the goal of bringing the residual spared sensorimotor system into a more excitable/relaxed state as a precursor for enhanced neural plasticity and accelerated recovery.

Although the sensorimotor circuit can be potentiated through BCI-neurofeedback paradigms (Ang and Guan 2017; Ramos-Murguialday et al. 2013; Soekadar et al. 2015a), less is known about whether such BCI-neurofeedback can explicitly guide sensorimotor cortical activation to a targeted hemisphere. This is crucial to understanding the brain's intrinsic neural interferences (e.g., the interaction between the left and right hemispheres through transcallosal connections, and interhemispheric connectivity of the supplementary motor area-SM1 network) (Arai et al. 2011; Waters et al. 2017). Sensorimotor circuits in left and right hemispheres might potentially influence one another, suggesting that BCI-neurofeedback of the SMR signal from one hemisphere does not always guarantee spatially specific activation of the sensorimotor circuit in the targeted hemisphere (Buch et al. 2008; Caria et al. 2011). Indeed, no previous study has shown that the sensorimotor cortical activity can be guided to the targeted hemisphere with spatial specificity, either contralaterally or ipsilaterally.

Guiding spatially specific sensorimotor cortical activation is important, for example, for facilitating neural rehabilitation processes. For instance, the remodeling process of the ipsilesional SM1 for finger motor recovery in the post-stroke stage was shown to impede shoulder movement recovery, as the enlargement of motor areas associated with finger control can lead to erosion of motor areas responsible for shoulder movement (Muellbacher et al. 2002). Furthermore, contralesional SM1 demonstrated better control of paralyzed muscles than did ipsilesional SM1 (Takasaki 2017; McPherson et al. 2018). Thus, preventing competitive reinnervation processes in the SM1 may aid in better motor improvement post-stroke. BCI-neurofeedback designed to guide sensorimotor cortical activation to a targeted hemisphere has great potential to facilitate the neural remodeling process during post-stroke rehabilitation.

To resolve this uncertainty in the BCI-neurofeedback technique—that is, whether the sensorimotor cortical activity can be guided to the targeted hemisphere—we conducted a BCI experiment focusing on the neuroanatomical properties of skeletal muscle innervation as a pre-clinical trial and a First-in-Person Proof-of-Concept study. Recent studies suggested that there is a relationship between intrinsic functional/structural architecture of the brain and successful learning of brain activity (Halder et al. 2013; Young et al. 2016). In the present study, we selected two different motor imageries (MIs): "shoulder" MI in a first setting and "hand" MI in a second setting as a negative control. It is known that the deltoid anterior (DA) muscle for flexing proximal muscles is innervated bilaterally (Carson 2005; Colebatch et al. 1990). Conversely, the extensor digitorum communis (EDC) muscle, which is for extending distal muscles and is predominantly innervated from the contralateral hemisphere (Carson 2005; Colebatch et al. 1990), was used as a contrast to the bilateral corticomuscular connections of the DA muscle. Therefore, we hypothesized that, if BCI-neurofeedback is a potent up-regulator of hemispheric activation to the targeted side, shoulder MI-associated BCI-neurofeedback should enable the sensorimotor excitability to be lateralized to the targeted hemisphere, either contralaterally or ipsilaterally. In contrast, hand MI-associated BCI-neurofeedback might enable the sensorimotor excitability to be lateralized to the contralateral hemisphere, while limiting lateralization of the ipsilateral excitability by virtue of its neuroanatomical constraint.

In this study, participants performed shoulder/hand MI-associated BCI-neurofeedback to learn volitional regulation of sensorimotor cortical excitability in the contralateral or ipsilateral hemisphere in a double-blind, randomized, within-subject crossover design. We used a new BCI-neurofeedback approach during unilateral repetitive kinesthetic MI to volitionally regulate sensorimotor cortical excitability, as reflected by desynchronization/synchronization of SMR signals (SMR-ERD/ERS), with the aim of guiding its intensity to only the targeted hemisphere. To this end, we designed BCI-neurofeedback that displays both left and right hemispheric SMR-ERDs concurrently, allowing participants to learn to regulate these two variates at the same time and to modulate target-hemisphere-dependent SMR-ERD. This neuroanatomically-inspired approach enabled us to investigate potent neural remodeling functions that underlie EEG oscillation-based neurofeedback via a BCI.

## 2 Spatially Bivariate EEG-Based Neurofeedback

All participants completed the four different neurofeedback sessions on separate days; each session consisted of the pre- and post-evaluation blocks and the six training blocks (Fig. 1).

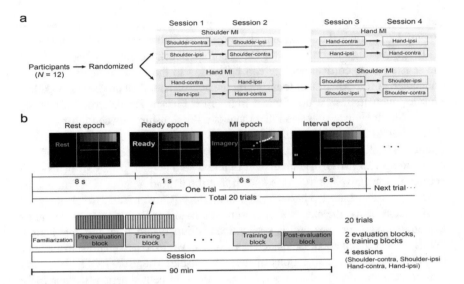

**Fig. 1** Study design and experimental paradigm

## 2.1 Evaluation Block

In the pre- and post-evaluation blocks, no visual feedback was provided. The aim of the pre-evaluation block was to evaluate the baseline brain activity and to calibrate parameters in the neurofeedback settings each day. First, the target frequency was calibrated for each participant in order to feedback the most reactive frequency. The target frequency was selected from the alpha band (8–13 Hz) by calculating the mean intensity of SMR-ERD with a 3-Hz sliding bin and 2-Hz overlap. SMR-ERD in the alpha band is a reliable EEG biomarker representing increased neuronal excitability in SM1, corticospinal tract, and thalamocortical systems. Second, the target level of SMR-ERD during MI was normalized for each participant at the third quartile of the contralateral or ipsilateral SMR-ERD in the pre-evaluation block. This setting was empirically approved by the authors as a moderate load for effective operant learning (Naros et al. 2016).

## 2.2 Training Block

In the training blocks, participants received visual feedback based on the SMR-ERDs from both left and right hemispheres. The real-time SMR-ERD intensity in each hemisphere (relative to the average power of the last 6 s of the resting epoch) was obtained every 100 ms. To modulate target hemisphere-dependent SMR-ERD, we developed BCI-neurofeedback that displayed both left and right hemispheric SMR-ERDs concurrently, allowing participants to learn to regulate these two variates at the same time. Visual feedback was provided on a computer screen in the form of cursor movements in a two-dimensional coordinate, in which each axis corresponded to the degree of the contralateral or ipsilateral SMR-ERD (Fig. 1). The axis range was from −100% (i.e., ERS) to 100%, and the cursors were presented at the origin-position $(x = 0, y = 0)$ at the initiation of a trial. A key point of this study is that participants were always instructed to try to move the cursor toward the upper right in the two-dimensional coordinate during MI in all four neurofeedback training sessions. In the case of shoulder MI, for example, participants performed the same MI and tried to move the cursor to the upper right regardless of whether it was a Shoulder-contra or Shoulder-ipsi session. However, the coordinate systems during the two sessions differed as follows: in the Shoulder-contra session, the x-axis and y-axis corresponded to the ipsilateral SMR-ERS and contralateral SMR-ERD, respectively. Conversely, in the Shoulder-ipsi session, the x-axis and y-axis corresponded to the contralateral SMR-ERS and ipsilateral SMR-ERD, respectively. Thus, the upper right position always indicated a reduction in alpha rhythm in the targeted hemisphere with respect to the baseline (i.e., SMR-ERD) and an increase in the non-targeted side (i.e., SMR-ERS). Using such a gimmicked environment, we aimed at lateralizing cortical activity in the sensorimotor cortex, blinding which task was being performed.

A score was calculated when the most recent cursor on the screen reached the ten blue boxes representing the scoring range (Fig. 1b). The coordinates of each blue box corresponded to the degree of bilateral SMR-ERDs, with the x-axis set in steps of 10% SMR-ERS in the non-targeted hemisphere, and y-axis ranged from the predefined threshold to 100% SMR-ERD in the target hemisphere. At the end of the trial, the computer cursor returned to the origin position. A score for each segment (each computer cursor updated every 100 ms) was obtained during the MI epoch to provide feedback about the overall performance of each trial. The darkest blue box in the upper left in Fig. 1b had a low score (5 points), whereas the lightest blue box in the upper right had a high score (15 points). The boxes in the middle were set in steps of 1 point. Finally, a cumulative sum calculated by adding all scores was displayed for 5 s in the left side of the screen at the interval period in each trial (range: 0–765 points). To boost learning of self-regulation in sensorimotor cortical activity, participants were encouraged to get a higher cumulative sum than during the previous trial. Such screen presentation of the total score at the end of the trial is referred to as 'intermittent feedback' (Johnson et al. 2012). Previous studies demonstrated that providing intermittent feedback is a useful element for neurofeedback training (Shibata et al. 2011; Posse et al. 2003) because it probably reduces cognitive loads during MI.

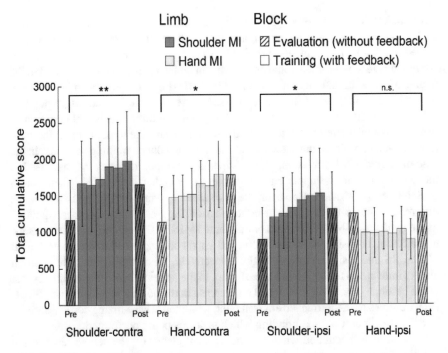

**Fig. 2** Changes in BCI performance

## 3 BCI Performance

Figure 2 illustrates the changes in BCI performance (i.e., total cumulative score) in each session. In shoulder MI, the total cumulative score was improved in both the Shoulder-contra session (pre = 1170, post = 1651, difference = 481, Cohen's $d$ = 0.72, $p$ = 0.007, paired t-test) and the Shoulder-ipsi session (pre = 891, post = 1310, difference = 419, Cohen's $d$ = 0.84, $p$ = 0.011, paired t-test). On the other hand, in hand MI, the total cumulative score was improved in the Hand-contra session (pre = 1783, post = 1136, difference = 647, Cohen's $d$ = 1.20, $p$ = 0.018, paired t-test), but not in the Hand-ipsi session (pre = 1249, post = 1253, difference = 4, Cohen's $d$ = 0.01, $p$ = 0.97, paired t-test). We also found that the total cumulative score in the evaluation blocks was lower than that in the training blocks, which is in keeping with the well-known information that visual feedback enhances MI-based BCI performance (Ono et al. 2015; Pichiorri et al. 2015). Differences between the 6 training blocks for cumulative score were not statistically significant (all $p$ > 0.05, one-way rmANOVA).

## 4 Effects of EEG-Based Neurofeedback During Shoulder MI

Spatial patterns of SMR-ERD during the MI epoch in the pre- and post-evaluation blocks of a representative participant are shown in Fig. 3a, b. The SMR-ERDs were localized predominantly in the bilateral parieto-temporal regions during the pre-evaluation block, regardless of whether it was a Shoulder-contra or Shoulder-ipsi session. During the Shoulder-contra session, the contralateral SMR-ERD increased after the neurofeedback training session, whereas the ipsilateral SMR-ERD did not (Fig. 3a). Conversely, during the Shoulder-ipsi session, the contralateral SMR-ERD did not increase, but the ipsilateral SMR-ERD did (Fig. 3b).

Figure 3c, d show Laterality Index (LI) changes during the Shoulder-contra and Shoulder-ipsi sessions, respectively. During the Shoulder-contra session, the LI in the post-evaluation block ($-0.113 \pm 0.072$) was significantly lower than that in the pre-evaluation block ($-0.030 \pm 0.089$) (difference $= 0.083$, Cohen's $d = 1.76$, $p = 0.023$, paired t-test; Fig. 3c). By contrast, during the Shoulder-ipsi session, the LI in the post-evaluation block ($0.017 \pm 0.103$) was significantly higher than that in the pre-evaluation block ($-0.067 \pm 0.103$) (difference $= 0.084$, Cohen's $d = 0.86$, $p = 0.039$, paired $t$-test; Fig. 3d). Target-hemisphere-dependent SMR-ERDs were modulated during both the Shoulder-contra and Shoulder-ipsi sessions, even though participants repeated the same MI under the neurofeedback setting with only a change in the rule of cursor movement (i.e., reversal of x-axis and y-axis).

We assessed seed-based corrected imaginary part of coherence (ciCOH during the resting-state in the pre- and post-evaluation blocks to evaluate interregional synchronization (i.e., functional connectivity). Figure 4a, b show significant *intra*hemispheric connections in each hemisphere of a representative participant. The number of significant connections in the contralateral hemisphere increased from the pre- to the post-

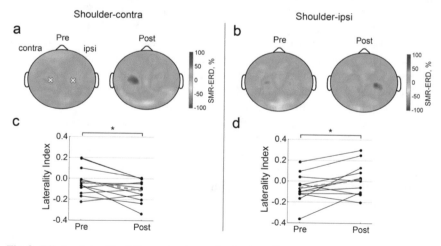

**Fig. 3** Effects of shoulder MI-associated neurofeedback on SMR-ERD

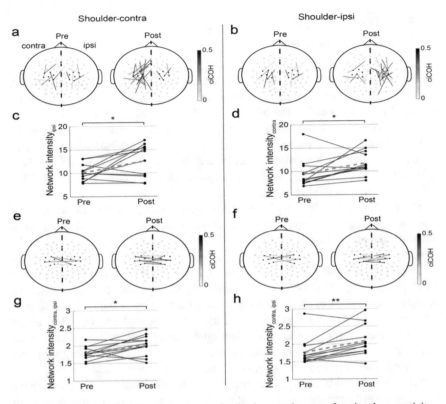

**Fig. 4** Effects of shoulder MI-associated neurofeedback on resting-state functional connectivity

epochs during the Shoulder-contra session (Fig. 4a), whereas they increased in the ipsilateral hemisphere during the Shoulder-ipsi session (Fig. 4b). *Intra*hemispheric network intensity changes in the targeted hemisphere during the Shoulder-contra and Shoulder-ipsi sessions are shown in Fig. 4c and d, respectively. Figure 4e, f show significant *inter*hemispheric connections of a representative participant, which increased during both the Shoulder-contra and Shoulder-ipsi sessions. Changes in interhemispheric network intensity for all participants during the Shoulder-contra and Shoulder-ipsi sessions are outlined in Fig. 4g and h, respectively. During the Shoulder-contra session, the interhemispheric network intensity was significantly higher during the post-evaluation block ($2.01 \pm 0.28$) than during the pre-evaluation block ($1.80 \pm 0.19$; difference $= 0.21$, Cohen's $d = 0.88$, $p = 0.030$, paired t-test; Fig. 4g). Similarly, during the Shoulder-ipsi session, the interhemispheric network intensity was significantly higher in the post-evaluation block ($2.09 \pm 0.42$) than in the pre-evaluation block ($1.77 \pm 0.37$) (difference $= 0.31$, Cohen's $d = 0.81$, $p = 0.006$, paired t-test; Fig. 4h).

## 5 Comparison of SMR-ERDs During Shoulder MI and Hand MI

To further examine the effectiveness in BCI-neurofeedback training purported to lateralize sensorimotor cortical activities, we compared the changes in SMR-ERD during shoulder MI and hand MI (Fig. 5). A three-way ANOVA revealed a significant interaction between Session × Hemisphere × Limb ($F_{(1, 88)} = 4.98, p = 0.047$) and Session × Hemisphere ($F_{(1, 88)} = 26.7, p < 0.001$), but no interaction between Session × Limb ($F_{(1, 88)} = 1.44, p = 0.255$) or Hemisphere × Limb ($F_{(1, 88)} = 2.06 p = 0.179$). Although Limb had a significant main effect ($F_{(1, 88)} = 5.43, p = 0.040$), Session ($F_{(1, 88)} = 1.29, p = 0.28$) and Hemisphere ($F_{(1, 88)} = 2.39, p = 0.15$) did not have any effects. Post hoc two-way ANOVA with Hemisphere × Limb in the sessions aiming for lateralization to the contralateral hemisphere (i.e., Shoulder-contra and Hand-contra) showed a significant main effect for Hemisphere ($F_{(1, 44)} = 12.39, p = 0.001$); however, there was no main effect for Limb ($F_{(1, 44)} = 0.27, p = 0.608$) and no interaction ($F_{(1, 44)} = 0.30, p = 0.589$). By contrast, post hoc two-way ANOVA with Hemisphere × Limb in the sessions aiming for lateralization to the ipsilateral hemisphere (i.e., Shoulder-ipsi and Hand-ipsi) indicated a significant main effect for Hemisphere ($F_{(1, 44)} = 4.51, p = 0.039$) and interaction ($F_{(1, 44)} = 5.70, p = 0.021$), but no main effect for Limb ($F_{(1, 44)} = 1.99, p = 0.166$). Thus, there were interhemispheric differences in $\Delta$SMR-ERD during the shoulder MI and hand MI tasks. Moreover, a post hoc paired $t$-test demonstrated a significant difference in Hemisphere ($p = 0.001$) during shoulder MI, but no difference in Hemisphere ($p = 0.859$) during hand MI (Bonferroni corrected). Thus, the $\Delta$SMR-ERD in the ipsilateral hemisphere was significantly more positive than that in the contralateral

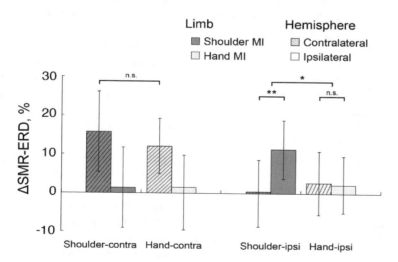

**Fig. 5** Two-way interaction in $\Delta$SMR-ERD during shoulder MI (dark gray) and hand MI (light gray) tasks

hemisphere during the shoulder MI task, but no significant difference was observed during the hand MI task.

## 6 Discussion

### 6.1 The Lateralization of Sensorimotor Cortical Activity to the Contralateral Hemisphere

The SMR-ERD contralateral to the imagined limb increased significantly after both the shoulder-contra and Hand-contra sessions. Previous studies also demonstrated up-conditioning of the contralateral SM1 using contralateral-based neurofeedback during hand MI (Birbaumer and Cohen 2007; Ang et al. 2011; Prasad et al. 2010). Repetitive induction of the SMR-ERD contralateral to the imagined limb through visual or sensory feedback with neuromuscular electrical stimulation or robotic movement supports are considered to induce the use-dependent, error-based, and/or Hebbian-like plasticity of the contralateral SM1 (Gharabaghi et al. 2014; Soekadar et al. 2015b; Ushiba and Soekadar 2016). As contralateral SMR-ERD is a surrogate monitoring marker of contralateral SM1 excitability (Takemi et al. 2013, 2015; Hayashi et al. 2019), BCI-neurofeedback can promote operant learning of contralateral sensorimotor cortical activity. This is an expected phenomenon because distal muscles such as the EDC muscle are innervated from the contralateral hemisphere, which is most influential for muscle contraction (Carson 2005; Colebatch et al. 1990).

However, the BCI-neurofeedback-derived SMR signal from the contralateral hemisphere does not always guarantee spatially specific activation of the contralateral SM1 because both hemispheres are connected by intrinsic transcallosal projections and exhibit functional crosstalk (Arai et al. 2011; Waters et al. 2017). Indeed, conventional contralateral-based BCI-neurofeedback has induced a global increase including the ipsilateral SMR-ERD, indicating conventional BCI is considered as a modulation technique without spatial specificity (Pichiorri et al. 2015; Birbaumer and Cohen 2007; Ono et al. 2014). A key advantage of our study was that the BCI-neurofeedback that we developed monitored both contralateral and ipsilateral SMR-ERDs, demonstrating explicitly guided sensorimotor cortical activation in the targeted contralateral hemisphere alone.

Guiding cortical sensorimotor activation to the targeted hemisphere is also crucial in the context of neurorehabilitation. For example, it is known that an imbalanced interhemispheric inhibition due to excessive suppression from the ipsilateral (contralesional) to the contralateral side results in further attenuation of the contralateral sensorimotor cortical activity (Shimizu et al. 2002; Murase et al. 2004; Bütefisch et al. 2008). A temporary guide to down-conditioning in the ipsilateral hemisphere using non-invasive brain stimulation is also important to reduce interhemispheric inhibition after stroke (Hummel and Cohen 2006; Takeuchi et al. 2012). Therefore, the laterality shifting of sensorimotor cortical activity to the contralateral side may contribute to the degree of achievable functional recovery (Askim et al. 2009; Chieffo et al. 2013).

## 6.2 The Lateralization of Sensorimotor Cortical Activity to the Ipsilateral Hemisphere

During the shoulder-ipsi session, the ipsilateral SMR-ERD increased significantly. Although increasing evidence suggests that the contribution of the ipsilateral hemisphere is salient in motor control (Ward et al. 2003; Dodd et al. 2017; Bundy et al. 2017), no previous study has shown that sensorimotor cortical activity can be guided to the ipsilateral hemisphere. Chiew and his colleagues indicated that different types of MI-based (right and left hands) fMRI neurofeedback of the LI (i.e., the difference in BOLD responses between the contralateral M1 and the ipsilateral M1 to the imagined hand) is capable of lateralizing to the contralateral hemisphere (Chiew et al. 2012), but lateralizing to the ipsilateral was not successful due to "hand" MI-associated neurofeedback. Therefore, our study is the first to show that BCI-neurofeedback is a potent up-regulator of hemispheric activation to the targeted hemisphere, either contralaterally or ipsilaterally in the same participants, depending on the targeted muscle.

Successful up-conditioning of the ipsilateral SM1 during shoulder MI may be associated with its neuroanatomical properties, because ipsilateral SMR-ERD reflects the excitability of the ipsilateral corticospinal tract (CST) (Hasegawa et al. 2017), which mainly innervates proximal muscles (Carson 2005; Alawieh et al. 2017). Unlike hand motor muscles, the functional recovery of axial or shoulder muscles following stroke hemiplegia is promoted by unmasking the ipsilateral pathway to the paretic hand (Muellbacher et al. 2002; Colebatch et al. 1990; Schwerin et al. 2008). Thus, neurofeedback aimed at ipsilateral lateralization would be conceptually useful for stroke rehabilitation, particularly for functional maturation of ipsilateral CST and proximal muscle motor recovery.

The ipsilateral SMR-ERD did not increase during the Hand-ipsi session, implying that the extent of corticospinal projection from the ipsilateral hemisphere to the imagined body part affected the modulation of the laterality of sensorimotor cortical activity. With recent developments in neuroimaging techniques, there is an emerging interest in understanding the intrinsic functional and structural architecture of the brain that underlies successful learning of brain activity. For example, research probing the prediction of BCI aptitude from individual brain structures demonstrated that the integrity and myelination quality of deep white matter structures, such as the corpus callosum, cingulum, and superior fronto-occipital fascicle, were positively correlated with individual BCI-performance (Halder et al. 2013). Additionally, it has been suggested that changes in the integrity of the contralesional CST may be accompanied by improved BCI-performance after stroke (Young et al. 2016). The current literature makes it clear that there is a relationship between neuroanatomical characteristics and voluntary control of brain activity. Therefore, our findings implied that intrinsic neuroanatomical properties such as the CST constrains the effectiveness in BCI-neurofeedback training purported to lateralize sensorimotor cortical activities. Further work that approaches the further understandings of differences in BCI-learning is warranted.

# 7 Conclusion

We addressed whether sensorimotor cortical activity can be guided to the targeted hemisphere using a BCI-neurofeedback approach that displays both left and right hemispheric SMR-ERDs to modulate bilateral sensorimotor cortical activities. EEG-based BCI-neurofeedback enabled us to up-regulate hemispheric activation to the targeted hemisphere, both contralaterally or ipsilaterally, in the same participants, which had not been reported prior to this work. During shoulder MI-associated neurofeedback, both the contralateral SMR-ERD and ipsilateral SMR-ERD increased. Network intensity in the targeted hemisphere also increased in association with increases in SMR-ERD, implying that the modulation of distributed interregional neural communication influenced the up-regulation of sensorimotor cortical activity. Conversely, the absence of an increase in ipsilateral SMR-ERD during hand MI indicated that the amount of corticospinal projection from the ipsilateral hemisphere to the imagined body part can constrain the laterality of brain activity. These results suggest that EEG-based BCI-neurofeedback that guides sensorimotor cortical activation in a targeted hemisphere, either contralaterally or ipsilaterally, has great potential to facilitate the resumption and shaping of the neural remodeling process.

# References

A. Alawieh, S. Tomlinson, D. Adkins, S. Kautz, W. Feng, Preclinical and clinical evidence on ipsilateral corticospinal projections: Implication for motor recovery. Transl. Stroke Res. **8**, 529–540 (2017)

K.K. Ang, C. Guan, EEG-based strategies to detect motor imagery for control and rehabilitation. IEEE Trans. Neural Syst. Rehabil. Eng. **25**, 392–401 (2017)

K.K. Ang et al., A large clinical study on the ability of stroke patients to use an EEG-based motor imagery brain-computer interface. Clin. EEG Neurosci. **42**, 253–258 (2011)

N. Arai et al., State-dependent and timing-dependent bidirectional associative plasticity in the human SMA-M1 network. J. Neurosci. **31**, 15376–15383 (2011)

T. Askim, B. Indredavik, T. Vangberg, A. Håberg, Motor network changes associated with successful motor skill relearning after acute ischemic stroke: a longitudinal functional magnetic resonance imaging study. Neurorehabil. Neural Repair **23**, 295–304 (2009)

N. Birbaumer, L.G. Cohen, Brain-computer interfaces: communication and restoration of movement in paralysis. J. Physiol. (Lond.) **579**, 621–636 (2007)

Buch et al., Think to move: a neuromagnetic Brain-Computer Interface (BCI) system for chronic stroke. Stroke **39**, 910–917 (2008)

Bundy et al., Contralesional brain-computer interface control of a powered exoskeleton for motor recovery in chronic stroke survivors. Stroke **48**, 1908–1915 (2017)

C.M. Bütefisch, M. Weβling, J. Netz, R.J. Seitz, V. Hömberg, Relationship between interhemispheric inhibition and motor cortex excitability in subacute stroke patients. Neurorehabil. Neural Repair **22**, 4–21 (2008)

A. Caria et al., Chronic stroke recovery after combined BCI training and physiotherapy: a case report. Psychophysiology **48**, 578–582 (2011)

R.G. Carson, Neural pathways mediating bilateral interactions between the upper limbs. Brain Res. Rev. **49**, 641–662 (2005)

R. Chieffo et al., Mapping early changes of cortical motor output after subcortical stroke: a transcranial magnetic stimulation study. Brain Stimul. **6**, 322–329 (2013)

M. Chiew, S.M. LaConte, S.J. Graham, Investigation of fMRI neurofeedback of differential primary motor cortex activity using kinesthetic motor imagery. NeuroImage **61**, 21–31 (2012)

J.G. Colebatch, J.C. Rothwell, B.L. Day, P.D. Thompson, C.D. Marsden, Cortical outflow to proximal arm muscles in man. Brain **113**, 1843–1856 (1990)

K.C. Dodd, V.A. Nair, V. Prabhakaran, Role of the contralesional vs. ipsilesional hemisphere in stroke recovery. Front. Hum. Neurosci. **11**, 9 (2017)

A. Gharabaghi et al., Coupling brain-machine interfaces with cortical stimulation for brain-state dependent stimulation: enhancing motor cortex excitability for neurorehabilitation. Front. Hum. Neurosci. **8**, 7 (2014)

S. Halder et al., Prediction of brain-computer interface aptitude from individual brain structure. Front. Hum. Neurosci. **7**, 9 (2013)

K. Hasegawa et al., Ipsilateral EEG mu rhythm reflects the excitability of uncrossed pathways projecting to shoulder muscles. J. NeuroEng. Rehabil. **14**, 11 (2017)

M. Hayashi et al., Two-stage regression of high-density scalp electroencephalograms visualizes force regulation signaling during muscle contraction. J. Neural Eng. **16**, 056020 (2019)

F.C. Hummel, L.G. Cohen, Non-invasive brain stimulation: a new strategy to improve neurorehabilitation after stroke? Lancet Neurol. **5**, 708–712 (2006)

K.A. Johnson et al., Intermittent 'real-time' fMRI feedback is superior to continuous presentation for a motor imagery task: a pilot study. J. Neuroimaging **22**, 58–66 (2012)

J.G. McPherson et al., Progressive recruitment of contralesional cortico-reticulospinal pathways drives motor impairment post stroke. J. Physiol. **596**, 1211–1225 (2018)

W. Muellbacher et al., Improving hand function in chronic stroke. Arch. Neurol. **59**, 1278–1282 (2002)

N. Murase, J. Duque, R. Mazzocchio, L.G. Cohen, Influence of interhemispheric interactions on motor function in chronic stroke. Ann. Neurol. **55**, 400–409 (2004)

G. Naros, I. Naros, F. Grimm, U. Ziemann, A. Gharabaghi, Reinforcement learning of self-regulated sensorimotor β-oscillations improves motor performance. NeuroImage **134**, 142–152 (2016)

C. Neuper, G. Pfurtscheller, Event-related dynamics of cortical rhythms: frequency-specific features and functional correlates. Int. J. Psychophysiol. **43**, 41–58 (2001)

C. Neuper, M. Wörtz, G. Pfurtscheller, ERD/ERS patterns reflecting sensorimotor activation and deactivation. Prog. Brain Res. **159**, 211–222 (2006)

T. Ono et al., Brain-computer interface with somatosensory feedback improves functional recovery from severe hemiplegia due to chronic stroke. Front. Neuroeng. **7**, 19 (2014)

T. Ono et al., Multimodal sensory feedback associated with motor attempts alters BOLD responses to paralyzed hand movement in chronic stroke patients. Brain Topogr. **28**, 340–351 (2015)

G. Pfurtscheller, C. Brunner, A. Schlögl, F.H. Lopes da Silva, Mu rhythm (de)synchronization and EEG single-trial classification of different motor imagery tasks. Neuroimage **31**, 153–159 (2006)

F. Pichiorri et al., Brain-computer interface boosts motor imagery practice during stroke recovery. Ann. Neurol. **77**, 851–865 (2015)

S. Posse et al., Real-time fMRI of temporolimbic regions detects amygdala activation during single-trial self-induced sadness. NeuroImage **18**, 760–768 (2003)

G. Prasad, P. Herman, D. Coyle, S. McDonough, J. Crosbie, Applying a brain-computer interface to support motor imagery practice in people with stroke for upper limb recovery: a feasibility study. J. Neuroeng. Rehabil. **7**, 60 (2010)

A. Ramos-Murguialday et al., Brain-machine-interface in chronic stroke rehabilitation: a controlled study. Ann. Neurol. **74**, 100–108 (2013)

S. Schwerin et al., Ipsilateral versus contralateral cortical motor projections to a shoulder adductor in chronic hemiparetic stroke: implications for the expression of arm synergies. Exp. Brain Res. **185**, 509–519 (2008)

K. Shibata, T. Watanabe, Y. Sasaki, M. Kawato, Perceptual learning incepted by decoded fMRI neurofeedback without stimulus presentation. Science **334**, 1413–1415 (2011)

T. Shimizu et al., Motor cortical disinhibition in the unaffected hemisphere after unilateral cortical stroke. Brain **125**, 1896–1907 (2002)

S.R. Soekadar, M. Witkowski, N. Birbaumer, L.G. Cohen, Enhancing Hebbian learning to control brain oscillatory activity. Cereb. Cortex **25**, 2409–2415 (2015a)

S.R. Soekadar, N. Birbaumer, M.W. Slutzky, L.G. Cohen, Brain–machine interfaces in neurorehabilitation of stroke. Neurobiol. Dis. **83**, 8 (2015b)

K. Takasaki, Targeted up-conditioning of contralesional corticospinal pathways promotes motor recovery in poststroke patients with severe chronic hemiplegia. *The Annual BCI Award 2017; The 12 Nominees* (2017)

M. Takemi, Y. Masakado, M. Liu, J. Ushiba, Event-related desynchronization reflects downregulation of intracortical inhibition in human primary motor cortex. J. Neurophysiol. **110**, 1158–1166 (2013)

M. Takemi, Y. Masakado, M. Liu, J. Ushiba, Sensorimotor event-related desynchronization represents the excitability of human spinal motoneurons. Neuroscience **297**, 58–67 (2015)

N. Takeuchi, Y. Oouchida, S.-I. Izumi, Motor control and neural plasticity through interhemispheric interactions. Neural Plast. **6**, 13 (2012)

J. Ushiba, S.R. Soekadar, Brain–machine interfaces for rehabilitation of poststroke hemiplegia. Prog. Brain Res. **228**, 163–183 (2016)

N.S. Ward, M.M. Brown, A.J. Thompson, R.S.J. Frackowiak, Neural correlates of outcome after stroke: a cross-sectional fMRI study. Brain **126**, 1430–1448 (2003)

S. Waters, T. Wiestler, J. Diedrichsen, Cooperation not competition: bihemispheric tDCS and fMRI show role for ipsilateral hemisphere in motor learning. J. Neurosci. **37**, 7500–7512 (2017)

B.M. Young et al., Brain-computer interface training after stroke affects patterns of brain-behavior relationships in corticospinal motor fibers. Front. Hum. Neurosci. **10**, 13 (2016)

# Next Generation Microscale Wireless Implant System for High-Density, Multi-areal, Closed-Loop Brain Computer Interfaces

Farah Laiwalla, Vincent W. Leung, Jihun Lee, Patrick Mercier,
Peter Asbeck, Ramesh Rao, Lawrence Larson, and Arto Nurmikko

**Abstract** A major challenge to high-resolution, closed-loop Brain Computer Interfaces (BCIs) is the availability of implantable technologies facilitating vastly parallel, large-scale access to cortical neural data representing complex, naturalistic tasks or sophisticated therapeutic neuromodulation. The current technological bottleneck is scalability of systems employing intra or epicortical electrode arrays with hard-wired tethers and bulky implant packaging. We address these challenges by employing an approach relying on spatially-distributed, completely wireless clusters of autonomous microscale neural interfaces, where each microdevice provides a single bidirectional channel (read-out and write-in) of neural access, and occupies a volume <0.01 mm$^2$ inclusive of biocompatible packaging for long-term implantation. Wireless power transfer, high-bandwidth bidirectional telecommunications and adaptive networking across multi-areal clusters are managed by a wearable external module to produce an implantable device system with anatomic flexibility and scalability, forming a "cortical internet".

**Keywords** Brain-Computer Interface (BCI) · Electrocorticographic (ECoG) ·
Bidirectional wireless neural interfaces · Neuroprosthetics · Cortical internet

## 1 Introduction

Modern high-performance Brain Computer Interfaces (BCIs) rely on high fidelity sensors in the form of implanted multichannel electrocorticographic (ECoG) or microelectrode arrays (MEAs). Several other chapters in this book demonstrate

F. Laiwalla (✉) · J. Lee · L. Larson · A. Nurmikko
Brown University School of Engineering, Providence, RI 02912, USA
e-mail: farah_laiwalla@brown.edu

V. W. Leung · P. Mercier · P. Asbeck · R. Rao
University of California San Diego, La Jolla, CA 92093, USA

V. W. Leung
Baylor University, Waco, TX 76798, USA

© The Author(s), under exclusive license to Springer Nature Switzerland AG 2021
C. Guger et al. (eds.), *Brain-Computer Interface Research*,
SpringerBriefs in Electrical and Computer Engineering,
https://doi.org/10.1007/978-3-030-60460-8_4

systems built upon these types of sensor technologies, which offer 100–200 channels of neural access, typically through surgically implanted sensors with percutaneous wired connections. While these state-of-the-art neural interface systems provide adequate neural access to reliably decode for low-dimensional tasks such as the constrained 2-D control of a computer cursor, it is anticipated that complex movements would need significantly higher channel counts which may be incompatible with the current monolithic, tethered implant platforms. Meanwhile, recent progress in integrated microtechnologies and wireless communications has culminated in the development of the first generation of wireless implantable neural interfaces (Gao et al. 2012; Borton et al. 2013; Yin et al. 2013). Although these devices rely on the same sensor front-end, their electronic integration has enabled, for example, research in freely moving monkeys (Borton et al. 2013), as well as human pilot trials (Simeral et al. 2019).

The next frontier in BCI technologies is to enable a large extension in the channel count capabilities of implanted neural interfaces. Researchers at Columbia have proposed a >65000 channel monolithic system leveraging CMOS camera design techniques (Tsai et al. 2017), while the Neuropixel device developed at Janelia Research Labs (Jun et al. 2017) has provided researchers an avenue to simultaneously access multiple cortical depths in their >1000 channel device (Juavinett et al. 2019; Dutta et al. 2019). The notion of a dense, vastly parallel neural implant immediately raises the question of implant approaches for such a device, and researchers at Neuralink, Inc. (Musk 2019), have offered one possible solution through their demonstration of a robot-assisted surgical placement approach.

These valiant endeavors represent the cutting edge of technology. Yet they raise critical questions regarding a system level approach enabling the deployment of such large-scale implants, most notably the associated "implant burden" from the footprint of a large monolithic device, as well as the challenges in managing the placement, tethering and data transmission bandwidths. The view in our team is that from a scalability perspective, a dynamic form factor and distributed deployment are desirable—and the latter is critically tied to the capability for a robust wireless communication link. This chapter describes our early development and validation of an approach that addresses current technological bottlenecks by implementing a distributed wireless microscale sensor system for neural interfaces.

## 2   System Architecture

It has recently been proposed that one way to achieve thousands or tens of thousands of parallel channels of neural access is to develop small (sub-mm), autonomous neural interfacing nodes, which may be distributed across the cortex (Seo et al. 2013; Yeon et al. 2016; Khalifa et al. 2018; Ahmadi et al. 2019). A distributed sensor system is conceptually promising, so long as the individual sensor node is ultraminiaturized to minimize volume overhead while maintaining a high-fidelity neural interface. It is then possible to envision a "cortical internet", comprising spatially-distributed clusters of sensors and actuators, providing bidirectional access to cortical neural networks from widespread brain areas in an adaptive fashion. We describe the development and testing validation from an epi-cortical implementation of this architecture, which we refer to as the "neurograin" system. To the authors' best knowledge at the time of this writing, this represents the first-ever reported demonstration of a completely wireless, synchronized, configurable network of sensors and microstimulators in a neural application.

The neurograin system constitutes ensembles of implantable, sub-millimeter, individually addressable, microelectronic chiplets. As shown in Fig. 1, each neurograin is designed as a self-contained, hermetically sealed module measuring 500 μm × 500 μm × 35 μm. For the initial prototype, we have chosen to implement a 1000-channel system, with an overall system latency of <100 ms (compatible with neural prosthetic applications). Design specifications such as uplink data rates (10 Mbps), downlink data rates (1 Mbps), packet duration (100 μs per channel) and packet periodicity (100 ms data frame) are derived from these considerations.

The neurograin device contains, at its core, a custom integrated chip (ASIC). This chip implements a radio frequency (RF) micro-antenna and all the associated sensor, stimulator, communication and networking circuits. The chip can be powered through transcutaneous wireless power delivery via near-field inductive coupling at approximately 1 GHz. The harvested energy activates on-chip analog and digital circuits responsible for neural signal recording and/or micro-stimulation.

For ultra-low-power uplink communication, BPSK-modulated RF backscattering has been employed. On the other hand, to accomplish robust downlink communication to free-floating implants without synchronous clocks and voltage references, a novel Amplitude Shift Keying Pulse Width Modulation (ASK-PWM) scheme has been designed. As shown in Fig. 2, neurograins' operations can be adaptively managed by an external wireless hub known as the "Epidermal Skinpatch" to form a time-domain multiple-access (TDMA) network. This external radio module, is software configurable (software defined radio or SDR), and has capabilities to leverage

**Fig. 1** Neurograins on a US penny; Photomicrograph of a recording neurograin chiplet

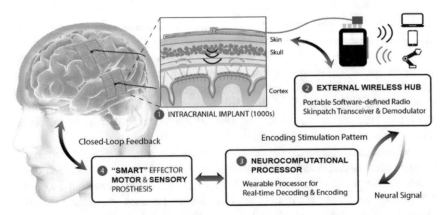

**Fig. 2** Concept of spatially-distributed implanted wireless neurograin read-out-write-in network with a wireless intra-cranial implant interfacing with an external wearable wireless hub, routing neural information to a real-time computational processor to drive feedback in a closed-loop BCI

an integrated high-performance Field-Programmable Gate Array (FPGA) to provide a host of modulation/demodulation functions in situ and in real-time. It thus provides the gateway to neurocomputational processing units enabling prosthetic control as well as neural encoding through patterned cortical stimulation.

Our network communication protocol is currently designed to accommodate up to 1000 channels of broadband ECoG data in this prototype system, but extendable via parallelization and other techniques to 10–100 × higher channel counts with modest hardware changes. Neurograin microdevices may be implanted individually or in ensembles (the latter as part of an ECoG grid or peripheral nerve cuff electrode, for example) into selected areas of the brain or peripheral nervous system (PNS), forming untethered, broadband, bidirectional neural interfacing elements usable for a variety of diagnostic and therapeutic neural applications.

An example epicortical (ECoG) sensor system is shown in Fig. 3, where a large number of untethered neurograins are organized into a 2-D grid that freely floats in the cerebrospinal fluid (CSF) surrounding the brain. A 3-coil electromagnetic coupling system (skinpatch transmit, relay and neurograin receive microantenna) have been designed to concentrate electromagnetic flux and reduce RF attenuation across 1 cm of biological tissues.

## 3   Neural Sensors and Stimulators

Recording neural activity with high signal to noise ratio (SNR) is a fundamental requirement for the wireless neurograin sensors. We have designed a self-standing, DC-coupled analog-front-end (AFE) with a merged amp-ADC architecture (Huang et al. 2018). In addition to conventional noise-reduction through chopping, we have

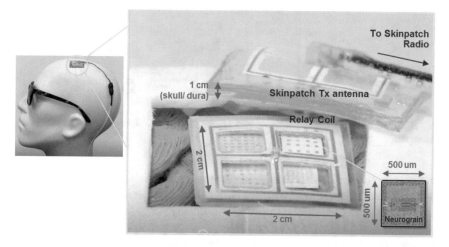

**Fig. 3** Cortical Surface (ECoG) Neurograin Sensor System. Individual neurograin (0.5 mm × 0.5 mm shown in inset) chips are embedded in a substrate to form a 2-D grid. Coupling between the chip micro antennas and the transmit "Skinpatch" antenna is optimized by the relay antenna

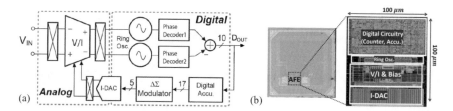

**Fig. 4** **a** Block diagram of the ultra-compact, low-noise AFE, and **b** IC micrograph of the AFE integrated into a wireless recording neurograin

implemented a VCO-based ADC and a mixed-signal differential electrode offset (DEO) cancellation servo loop (Fig. 4a) to eliminate capacitor use, which has led both to large area savings as well as elimination of kT/C noise. The AFE, which is designed for ECoG signals, has a bandwidth of 500 Hz and a dynamic range of ± 1 mV (where the latter is adequate for differential signals from neurograin electrodes with an inter-electrode spacing of ~100 μm). The electrode-interface DC-offset cancellation range is ± 50 mV. Input-referred noise is 2.2 μVrms over 500 Hz bandwidth for this implementation, and the total power consumption is 3.2 μW from a 0.6 V supply. Figure 4b shows a photomicrograph of the recording neurograin highlighting the AFE, which occupies 0.01 mm².

A programmable, charge-balanced biphasic current source is another required building block for bidirectional neural interfaces. The key specification for this current source is high voltage compliance, which would enable stimulation charge delivery across a range of tissue impedances. We have implemented a voltage-controlled resistor (VCR) based current-steering DAC as part of a source-reuse

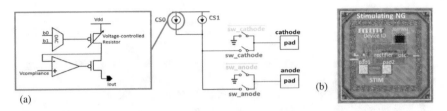

**Fig. 5** **a** Schematic of the bi-phasic current source, and **b** IC micrograph of the stimulating neurograin

architecture (Laiwalla et al. 2019). This scheme utilizes a single current source with changing directionality to provide for both anodic and cathodic phases of a stimulation pulse, thus providing intrinsic charge balance (Fig. 5a). The programmable neurograin current source provides control over current amplitude (up to 25 μA), pulse width (nominally 100 μs) and stimulation frequency (single pulse vs 100 Hz pulse train) through the wireless communication link. Figure 5b shows the IC micrograph of the programmable current source integrated into a wireless stimulating neurograin.

## 4   Wireless Power Transfer and Telecommunication

The neurograin implants are wirelessly powered in the near-field inductive coupling regime at 915 MHz. This is to limit tissue absorption while benefiting from an RF wavelength short enough to power sub-mm size micro-coils. The power budget for an individual neurograin is ~40 μW, and this must be harvested across a transcranial distance of ~1 cm. We have designed a multi-coil wireless power transfer (WPT) system, in this case introducing a relay antenna with moderate coupling with both the head-mounted transmitter and the implanted neurograin microreceiver antennas. This stacked 3-coil system increases the efficiency of WPT at least 50-fold in comparison with a conventional 2-coil approach. In addition, the transmit and relay antennas are designed as 3-D structures which are electromagnetically optimized to yield a uniform magnetic field over a large plane. This facilitates wireless access to a large number of spatially distributed microdevices placed within the area defined by the antenna perimeter. Figure 6 demonstrates one example of this multi-coil system, with a 4-quadrant coil design covering 2 cm × 2 cm and providing simultaneous wireless power and communication to 1000 neurograins as part of an anticipated human implant (Lee et al. 2018). Furthermore, these design concepts may be adapted based on anatomic considerations to be compatible with various animal models, with one example for a rodent system shown in Fig. 7.

The neurograin system incorporates both wireless power and bidirectional networked telecommunications into a single RF link (at ~1 GHz using the 3-coil antenna system described above). Figure 8a describes the chip-level implementation

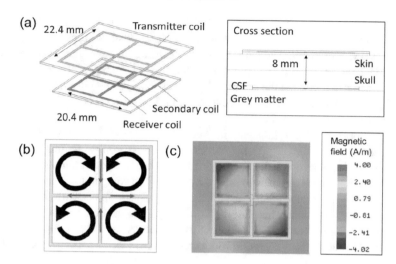

**Fig. 6** Transmit and relay antenna design **a** as area-matched 4-quadrant coils separated by 8 mm of tissue with **b** characteristic current flows and **c** uniform magnetic field strengths in each quadrant

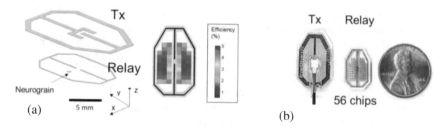

**Fig. 7** **a** Transmit and relay coil geometries and power-transfer efficiencies for a rodent-coil-system; and **b** Rodent implant and coil-system

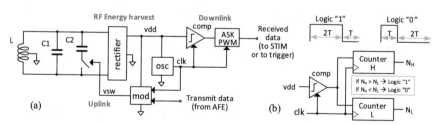

**Fig. 8** **a** Block diagram of the Neurograin RF energy harvesting, uplink and downlink circuits; and **b** the ASK-PWM scheme

of our approach, where a rectifier converts the coupled RF power to a DC supply voltage, powering all on-chip circuits. This also starts up a free-running oscillator which generates a ~30 MHz clock for on-chip digital logic. Neural data are sensed and sampled at 1 kilo Samples per second (kSa/s); 100 ms of data are held in an on-chip buffer, and packetized for uplink communication at 10 Mbps over 100 μs as per the TDMA protocol. The neurograin on-chip clock is used to convert the data into a Manchester-coded BPSK-encoded signal, which is then used to drive a switchable capacitor to modulate the reflected RF waves (backscatter) in a data-dependent way. This establishes a robust, ultra-low power uplink, with measured bit-error-rates (BER) <1e$^{-4}$ (Leung et al. 2018).

The skinpatch telecommunications hub is in charge of synchronization and scheduling across the entire neurograin network through issuing downlink commands. This is accomplished by varying transmitted RF power (amplitude modulation or ASK), which can then be sensed by an on-chip comparator. The neurograin clocks, however, are free-running and not frequency/phase aligned with each other or the downlink data stream. Therefore, data/frame synchronization cannot be achieved by simple ASK modulation. To address this challenge, we adopt a form of ASK-PWM (amplitude shift keying, pulse width modulation) scheme for the downlink data (Leung et al. 2019). This is described in Fig. 8b, where logic "1" and "0" are represented by high/low (H/L) pulse pairs with long/short (2T/T) and short/long (T/2T) durations respectively. Bits are thus encoded in the relative pulse width duration, and data synchronization is ensured by the low-to-high pulse transition, independent of the individual clock frequencies and phases of each neurograin. The ASK-PWM demodulation can be readily implemented with two digital counters and simple logic. For a downlink rate of 1 Mbps, we use a T of 0.33 μs (compared to a chip clock period of ~30 ns), and this allows adequate headroom to account for the ~20% variation in clock frequencies of different neurograins.

In order to arrange groups of neurograins into a network, each device is required to have a unique ID. We have implemented two different types of IDs into the neurograin devices. In Leung et al. (2019), 16-bit random addresses are achieved by area-efficient, low-power PUF (Physically Unclonable Function) circuits that leverage CMOS process variations to generate random-number IDs (Yang et al. 2017). In Laiwalla et al. (2019), 10-bit deterministic addresses are set by laser ablation of fuses realized with top layer redistribution metal during post-processing (Lee et al. 2020).

The external telecommunications hub is physically implemented using a wearable software-defined-radio, as shown in Fig. 9a. The latter has a field-programmable-

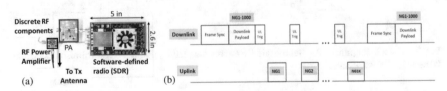

**Fig. 9** **a** external RF communication hub (Skinpatch), and **b** Timing diagram of the Skinpatch/neurograins TDMA network

gate-array (FPGA) back-end, which is utilized to provide real-time wireless demodulation, and has the computational capacity to serve as a processor for neural decoding/encoding for closed-loop control. This wearable module is ultimately anticipated to act as the bridge between the implant and remote computational resources, including cloud computing to leverage, for example, machine learning approaches to neural decoding.

We have demonstrated in our prototype devices, network-wide, synchronous downlink communication, and subsequent targeted (addressable) triggering of uplink backscattering in a "call-and-respond" manner. This forms a Time-Division Multiple Access (TDMA) network between the Skinpatch and a population of neurograins, with a timing-diagram shown in Fig. 9b. To the best of our knowledge, this work presents the first experimental validation of simultaneous wireless power transfer and Mbps bi-directional communications on a network of brain implant ICs over a single inductive coupling link.

## 5 Neurograin Post-processing and Packaging

Chronic biocompatibility of the neural implant is a consideration of utmost importance for clinical viability. A robust physical electrode-tissue interface is a major aspect of this challenge along with hermetic packaging of the implant. For the recording neurograins, we utilize standard post-process microfabrication techniques for patterned deposition of gold on top of the fabricated chiplet pads to form recording electrodes with impedances in the 100 k$\Omega$ range.

In contrast, stimulating neurograins require a lower electrode impedance for efficient charge transfer. For this purpose, we have investigated the integration of two types of biocompatible electrode interfaces: planar ECoG electrodes versus intracortically penetrating microwire electrodes. Both rely on direct post-process fabrication. For the planar electrodes, we have used poly(3,4-ethylenedioxythiophene) polystyrene sulfonate (PEDOT: PSS), an organic material with demonstrated history of use for neural stimulation applications (Ludwig et al. 2006). We have developed a photolithographic batch process to integrate 200 $\mu$m diameter PEDOT: PSS electrodes with the neurograin chiplets, while simultaneously embedding ensembles of devices in 25-$\mu$m thick liquid crystal polymer (LCP) sheets via thermocompression for hermetic encapsulation (Fig. 10b). For the penetrating electrode prototype, we have employed tungsten microwires with 75 $\mu$m diameter attached with conductive epoxy overlaid by insulating epoxy, as shown in Fig. 10c (Lee et al. 2020).

We have also concurrently developed techniques for ALD-based stacked multilayer conformal coatings for neurograin hermetic packaging (Jeong et al. 2019), as shown in Fig. 10a. The overall thickness profile of the packaging material is 100 nm, and the hermetically sealed chiplets have been tested in an accelerated aging testbed with demonstrated viability well over 10 years in an extrapolated physiological environment.

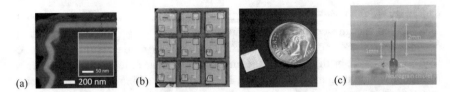

**Fig. 10** **a** Edge cross-section of a hermetically encapsulated neurograin, demonstrating stacked-multilayer ALD coatings, **b** nine-channel Stim neurograin ensemble, with 120 μm PEDOT electrodes, packaged via thermocompression between 25 μm thick LCP sheets, and **c** 3-D Stim neurograin with intracortical penetrating microwire electrodes

## 6  Validation

We have validated the full neurograin system, comprising ensembles of packaged microdevices and the wearable telecommunications hub in ex vivo rodent models enroute to the ongoing acute and chronic in vivo rodent implant trials. The recording and stimulation performance of epicortical neurograin arrays has been assessed in GFAP Cre x ChR2YFP mouse brain slices using a model for seizure induction (Lee et al. 2019). For this test, we overlaid 250 μm thick coronal cortical brain slices on top of an array of active neurograins in a modified standard immersion chamber, with the latter integrating the wireless telecommunication hardware (relay antenna). This is shown in Fig. 11a, while Fig. 11b highlights the raw wireless data streamed from a subset of active neurograins in the cortical area of interest. Baseline spontaneous field potential activity (LFPs) and response to electrical stimulation is captured,

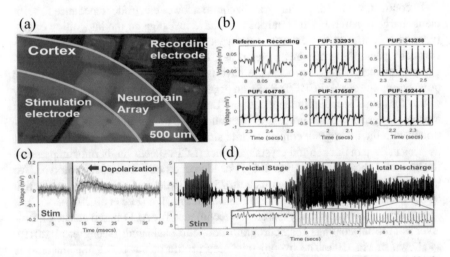

**Fig. 11** **a** Ex vivo neurograin experiment showing small neurograin ensemble under a coronal brain slice; **b** wireless data streaming from six neurograins visible in (a); **c** Characteristic stimulation-evoked depolarizations in cortical neurons, captured by neurograins; and **d** Induced epileptic seizure activity captured wirelessly by neurograin devices

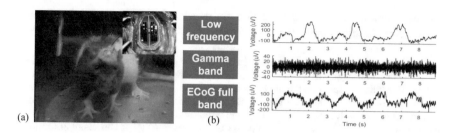

**Fig. 12  a** Neurograin implant in a freely moving rodent. Inset shows an ensemble of 56 neurograins, integrated with a custom relay antenna. **b** Intraop ECoG from a rodent using the 56-channel epicortical system in (a)

prior to applying chemical and electrical induction to produce epileptogenic activity. Recordings from the slices prior to seizure induction show characteristic slow LFP responses to electrical stimulation, which comprised 350 μA, 100 μs pulses at a frequency of 40 Hz over 500 ms (Fig. 11c). These post-stimulation depolarizations, lasting 10 to 20 ms, were recorded on multiple neurograins in a time-delayed manner representative of the spatial spread of the effects of stimulation pulses. Subsequently, we introduced the seizure inducing agent picrotoxin, which enables the capabilty to evoke seizure activity with triggered by electical stimulation. Preictal discharges and rhythmic bursts associated with the seizure activity were wirelessly recorded, as shown in Fig. 11d. These results validate the AFE's recording of physiological signals and TDMA networking fidelity over the long period of time.

The brain slice testbed was also utilized to validate the performance of stimulating neurograins. In this scenario, a hybrid network of stimulating and recording neurograins was used to trigger and measure responses. A 100 Hz, 25 μA Also symmetric pulse train was injected into the brain slices, and multiple stimulation triggered subthreshold depolarizations were captures on the recording electrode.

We are continuing to work toward providing a proof-of-concept validation for the neurograin system as a viable technology for chronic, implantable neural interfaces. We have built and tested a 56-channel rodent implant shown in Fig. 12a, and recorded wireless ECoG in an intra op setting using Ketamine anesthesia (Fig. 12b). Ongoing work continues to focus on system level optimizations to obtain wireless recordings from freely-moving and behaving rodents.

# 7  Summary

We present a system level implementation of a large scale, distributed wireless neural interface system comprising ensembles of autonomous neurograins, which integrate neural sensing and stimulation with wireless power delivery and bidirectional networked communications. Engineering design considerations and bench top and early animal testing and validation are presented in the context of building a robust, chronic implantable device system for early clinical translation.

**Acknowledgements** The work represented in this chapter is an interdisciplinary team effort, and the authors acknowledge with gratitude the many collaborators whose work is highlighted in this chapter. Regarding research on the implantable microdevices, we especially thank the following: Stefan Sigurdsson, Ah-Hyoung Lee, Huy Cu, Ethan Mok and Chester Kilfoyle at Brown University; Joonsoo Jeong (formerly at Brown University, and currently at Pusan National University, S. Korea); Jiannan Huang, Siwei Li, Siyuan Yu, Lingxiao Cui and Sravya Alluri at University of California San Diego, and Steven Shellhammer at Qualcomm. Initial phases of the research were supported by the DARPA NESD program.

# References

N. Ahmadi, M.L. Cavuto, P. Feng, L.B. Leene, M. Maslik, F. Mazza, et al., Towards a distributed, chronically-implantable neural interface. In *2019 9th International IEEE/EMBS Conference on Neural Engineering (NER)* (pp. 719–724). IEEE, Mar 2019

D.A. Borton, M. Yin, J. Aceros, A. Nurmikko, An implantable wireless neural interface for recording cortical circuit dynamics in moving primates. J. Neural Eng. **10**(2), 026010 (2013)

B. Dutta, A. Andrei, T.D. Harris, C.M. Lopez, J. O'Callahan, J. Putzeys, et al., The Neuropixels probe: A CMOS based integrated microsystems platform for neuroscience and brain-computer interfaces. In *2019 IEEE International Electron Devices Meeting (IEDM)* (pp. 10–11) (IEEE, 2019, Dec)

H. Gao, R.M. Walker, P. Nuyujukian, K.A. Makinwa, K.V. Shenoy, B. Murmann, T.H. Meng, HermesE: A 96-Channel full data rate direct neural interface in 0.13 $\mu$m CMOS. IEEE J. Solid-State Circuits **47**(4), 1043–1055 (2012)

J. Huang, F. Laiwalla, J. Lee, L. Cui, V. Leung, A. Nurmikko, P.P. Mercier, A 0.01-mm 2 mostly digital capacitor-less AFE for distributed autonomous neural sensor nodes. IEEE Solid-State Circuits Lett. **1**(7), 162–165 (2018)

J. Jeong, F. Laiwalla, J. Lee, R. Ritasalo, M. Pudas, L. Larson, et al., Conformal hermetic sealing of wireless microelectronic implantable chiplets by multilayered atomic layer deposition (ALD). Adv. Func. Mater. **29**(5), 1806440 (2019)

A.L. Juavinett, G. Bekheet, A.K. Churchland, Chronically implanted Neuropixels probes enable high-yield recordings in freely moving mice. Elife **8** (2019)

J.J. Jun, N.A. Steinmetz, J.H. Siegle, et al., Fully integrated silicon probes for high-density recording of neural activity. Nature **551**(7679), 232–236 (2017)

A. Khalifa, Y. Karimi, Q. Wang, S. Garikapati, W. Montlouis, M. Stanaćević, et al., The microbead: A highly miniaturized wirelessly powered implantable neural stimulating system. IEEE Trans. Biomed. Circuits Syst. **12**(3), 521–531 (2018)

F. Laiwalla, J. Lee, A.H. Lee, E. Mok, V. Leung, S. Shellhammer, et al. A distributed wireless network of implantable sub-mm cortical microstimulators for brain-computer interfaces. In *2019 41st Annual International Conference of the IEEE Engineering in Medicine and Biology Society (EMBC)* (pp. 6876–6879). IEEE, July 2019

J. Lee, F. Laiwalla, J. Jeong, C. Kilfoyle, L. Larson, A. Nurmikko, et al. Wireless power and data link for ensembles of sub-mm scale implantable sensors near 1 GHz. In *2018 IEEE Biomedical Circuits and Systems Conference (BioCAS)* (pp. 1–4). IEEE, Oct 2018

A.-H. Lee, J. Lee, F. Laiwalla, V. Leung, J. Huang, A. Nurmikko, Y.-K. Song, A scalable and low stress post-CMOS processing technique for Implantable neural Interfaces. *Submitted to Micromachines 2020*

J. Lee, E. Mok, J. Huang, L. Cui, A.H. Lee, V. Leung, et al., An implantable wireless network of distributed microscale sensors for neural applications. In *2019 9th International IEEE/EMBS Conference on Neural Engineering (NER)* (pp. 871–874). IEEE, Mar 2019

V.W. Leung, L. Cui, S. Alluri, J. Lee, J. Huang, E. Mok, et al., Distributed microscale brain implants with wireless power transfer and Mbps bi-directional networked communications. In *2019 IEEE Custom Integrated Circuits Conference (CICC)* (pp. 1–4). IEEE, April 2019

V.W. Leung, J. Lee, S. Li, S. Yu, C. Kilfovle, L. Larson, et al., A CMOS distributed sensor system for high-density wireless neural implants for brain-machine interfaces. In *ESSCIRC 2018-IEEE 44th European Solid-State Circuits Conference (ESSCIRC)* (pp. 230–233). IEEE, Sept 2018

K.A. Ludwig, J.D. Uram, J. Yang, D.C. Martin, D.R. Kipke, Chronic neural recordings using silicon microelectrode arrays electrochemically deposited with a poly (3, 4-ethylenedioxythiophene)(PEDOT) film. J. Neural Eng. **3**(1), 59 (2006)

E. Musk, An integrated brain-machine interface platform with thousands of channels. J. Med. Internet Res. **21**(10), e16194 (2019)

D. Seo, J.M. Carmena, J.M. Rabaey, E. Alon, M.M. Maharbiz, Neural dust: An ultrasonic, low power solution for chronic brain-machine interfaces. arXiv preprint (2013). arXiv:1307.2196

J.D. Simeral, T. Hosman, J. Saab, S.N. Flesher, M. Vilela, B. Franco, J. Kelemen, D.M. Brandman, J.G. Ciancibello, P.G. Rezaii, D.M. Rosler, K.V. Shenoy, J.M. Henderson, A.V. Nurmikko, L.R. Hochberg, Home use of a wireless intracortical brain-computer interface by individuals with Tetraplegia. medRxiv (2019)

D. Tsai, D. Sawyer, A. Bradd, R. Yuste, K.L. Shepard, A very large-scale microelectrode array for cellular-resolution electrophysiology. Nat. Commun. **8**(1), 1–11 (2017)

K. Yang, Q. Dong, D. Blaauw, D. Sylvester, 8.3 A 553F 2 2-transistor amplifier-based physically unclonable function (PUF) with 1.67% native instability. In *2017 IEEE International Solid-State Circuits Conference (ISSCC)* (pp. 146–147). IEEE, Feb 2017

P. Yeon, X. Tong, B. Lee, A. Mirbozorgi, B. Ash, H. Eckhardt, M. Ghovanloo, Toward a distributed free-floating wireless implantable neural recording system. In *2016 38th Annual International Conference of the IEEE Engineering in Medicine and Biology Society (EMBC)* (pp. 4495–4498). IEEE, Aug 2016

M. Yin, D.A. Borton, J. Aceros, W.R. Patterson, A.V. Nurmikko, A 100-channel hermetically sealed implantable device for chronic wireless neurosensing applications. IEEE Trans. Biomed. Circuits Syst. **7**(2), 115–128 (2013)

# Interfacing Hearing Implants with the Brain: Closing the Loop with Intracochlear Brain Recordings

**Ben Somers, Damien Lesenfants, Jonas Vanthornhout, Lien Decruy, Eline Verschueren, and Tom Francart**

**Abstract** The cochlear implant is one of the most successful rehabilitation prostheses, allowing deaf and severely hearing-impaired persons to hear again through electrical stimulation of the auditory nerve. In order to properly understand speech with a cochlear implant, a trained audiologist needs to adjust the sound processing and stimulation settings, which are highly subject-specific. Furthermore, this fitting procedure is time consuming, occurs only during infrequent visits to the clinic, and relies on behavioral feedback from the subject, which makes it challenging to do properly in young children and persons with cognitive impairment. Integrating a brain-computer interface (BCI) can alleviate the issues with the current fitting paradigms. If the implant can measure neural responses to speech, it can objectively assess how well the user understands speech and automatically adapt its sound processing settings if needed. This neuro-monitoring can happen continuously in the user's everyday listening environment and does not rely on behavioral input. We present an overview of our ongoing research towards such neuro-steered hearing implants.

**Keywords** Cochlear implant · Brain-computer interface · Neuro-steered hearing prostheses · Speech intelligibility · Electroencephalography (EEG)

## 1 Introduction

The cochlear implant (CI) is one of the most successful man-made interfaces to the neural system and is widely considered to be the most successful rehabilitation tool for severely hearing-impaired persons. A CI system consists of two parts: an external behind-the-ear piece and the implant itself (Dorman and Wilson 2004; Zeng et al. 2008) (see Fig. 1). The behind-the-ear piece is equipped with a microphone, sound processor, and battery. Sound captured by the microphone is converted into a sequence of electrical pulses and wirelessly transmitted to the implant. The internal

B. Somers (✉) · D. Lesenfants · J. Vanthornhout · L. Decruy · E. Verschueren · T. Francart
Department of Neurosciences, ExpORL, KU Leuven—University of Leuven,
Louvain, Belgium
e-mail: ben.somers@med.kuleuven.be

C. Guger et al. (eds.), *Brain-Computer Interface Research*,
SpringerBriefs in Electrical and Computer Engineering,
https://doi.org/10.1007/978-3-030-60460-8_5

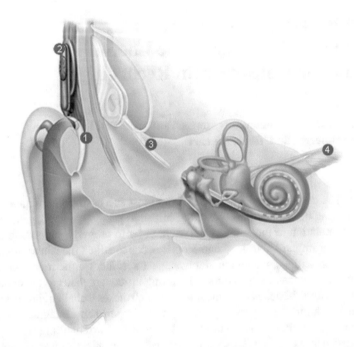

**Fig. 1** Cochlear implant system with external behind-the-ear piece (1, 2) and implant (3). A curved electrode array is inserted in the cochlea where it can electrically stimulate the auditory nerve (4). Copyright Cochlear Limited

part is surgically implanted, and contains a wireless receiver and an electrode array, which is inserted in the cochlea. The electrodes stimulate the various frequency-specific neural sites in the cochlea, restoring the user's ability to hear and even understand speech.

The conversion from sound to electrical stimulation involves many signal processing parameters (Wouters et al. 2015). After implantation, a specialized audiologist adjusts these parameters in the clinic, referred to as "fitting". A fitting is highly subject-specific, time consuming and labor intensive. Therefore, only a subset of stimulation settings can be adjusted during a fitting session. Furthermore, fittings only happen during infrequent visits at a clinic, in between which the CI operates as an open-loop system (see Fig. 2a) that does not consider variable factors that may affect the person's speech understanding, such as physiological changes, listening environment, or attentional state. The large variability in post-implantation outcomes (e.g. speech intelligibility) is partly caused by limitations of the fitting. As the current fitting paradigm is behavioral (i.e. active subject participation is required), performing a good fitting is even more challenging in certain clinical populations, such as young children, people with communication disorders, or elderly people with cognitive impairment.

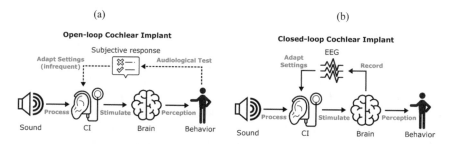

**Fig. 2** **a** Currently, CI settings are adjusted during infrequent visits to an audiologist, but the system operates as an open loop. Fitting sessions require behavioral feedback to the audiologist. **b** The envisioned closed-loop CI can continuously monitor the user's speech intelligibility and autonomously adapt its settings to improve performance. The system is objective as it does not rely on user behavior

In our research, we aim to overcome these issues by incorporating neural feedback in the implant, effectively introducing a continuous feedback mechanism into the system. In such a closed-loop brain computer interface, usually referred to as a "closed-loop CI" or "neuro-steered hearing prosthesis", brain responses to sound are recorded and processed in real-time to automatically and autonomously adapt the sound processing parameters for improving the user's speech understanding. This novel brain-computer interface would eliminate the need for frequent hospital visits and can operate without the need for behavioral feedback from the subject (Fig. 2b).

The realization of closed-loop hearing prostheses requires some fundamental scientific research. Firstly, as the main purpose of a CI is to restore speech understanding, an objective brain-based measure that reflects speech intelligibility is required (Sect. 2). Secondly, the simultaneous electrical stimulation of the CI interferes with the recordings of neural activity in the auditory pathway. Signal processing techniques to eliminate the effect of stimulation artifacts from neural recordings are required (Sect. 3). In a practical closed-loop system, the recording system should be integrated in the existing CI. We developed a technique to record neural activity from the implanted intracochlear electrodes (Sect. 4). Finally, algorithms need to be developed to adjust the stimulation settings of the CI based on the recorded objective measure (Sect. 5).

## 2 Objective Measures of Speech Intelligibility

The current state of the art method for measuring speech intelligibility is a behavioral test in which the subject repeats sentences from a standardized speech material, which are then manually scored by an audiologist. There exist clinical objective methods of hearing based on neural responses, but these generally indicate lower-level functioning of the auditory periphery. For instance, auditory brainstem responses can be used to measure detection thresholds, using short sound stimuli such as clicks, tone bursts or vowels. However, showing that a person can *hear* a sound is not the same as showing they can *understand* it.

The envelope of a speech signal is one of the most important features for speech intelligibility (Shannon et al. 1995). When listening to natural speech, neural oscillations in the brain track these slowly varying modulations of the speech signal (Aiken and Picton 2008; Peelle and Davis 2012). Neural recordings such as EEG or MEG, obtained while the subject is listening to continuous speech, can be used to reconstruct the original speech envelope (Ding and Simon 2012). In recent years, this "decoding" of the speech envelope from the listener's brain has been studied intensively, as the representation of the speech envelope in the brain provides insights on how we process and understand speech (Ding and Simon 2013).

We recently developed an EEG-based objective measure of speech intelligibility using natural speech in normal hearing listeners (Lesenfants et al. 2019; Vanthornhout et al. 2018; Verschueren et al. 2019b). A schematic overview of the method is depicted in Fig. 3a. The intelligibility of a speech stimulus is altered by adding stationary noise at different signal to noise ratios (SNRs). For reference, speech intelligibility is assessed behaviorally by having the subject repeat the sentences and scoring. The behavioral speech intelligibility score as a function of SNR follows a sigmoidal curve. The midpoint of this sigmoid is the SNR at which the subject understands 50% of the speech, and is referred to as the Speech Reception Threshold (SRT), a commonly

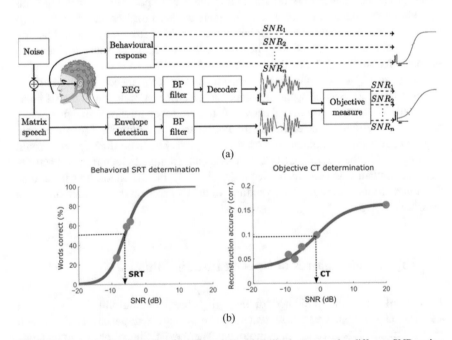

**Fig. 3** **a** Schematic overview of method. Speech intelligibility is measured at different SNRs using both a behavioral test (repeating sentences) and an objective measure (correlation between original and reconstructed speech envelope). **b** Derivation of behavioral SRT and objective CT measures as midpoint of sigmoidal curves. Figure adapted from Vanthornhout et al. (2018)

used clinical measure (Fig. 3b). For the objective measure, EEG is measured while the subject listens to the speech stimulus. After preprocessing such as bandpass filtering, artifact removal and noise reduction (Das et al. 2020; Somers et al. 2018), the speech envelope is reconstructed from the EEG with a linear decoder. By correlating it with the real envelope derived from the stimulus, we obtain a measure of neural tracking of the speech envelope as a function of SNR. This graph can be used to derive an objective measure of speech intelligibility referred to as the Correlation Threshold (CT), which was shown to correlate well with the clinical SRT (Vanthornhout et al. 2018).

The CT measure can be further improved in several ways. The neural tracking of the speech signal in the brain can be more accurately modeled by including complementary speech representations. A model integrating both low-level acoustical features such as the envelope and high-level information such as a phoneme representation obtains better results (Di Liberto et al. 2015; Di Liberto and Lalor 2016). By using such an integrated model combining both stimulus spectrogram and phonetic features, the objective CT measure was improved compared to the envelope-only based model (Lesenfants et al. 2019). For 80% of subjects, the difference between objective CT and behavioral, state-of-the art SRT was less than 2 dB, which is comparable to the measurement precision of the SRT (Decruy et al. 2018; Francart et al. 2010).

We demonstrated that the objective speech intelligibility measure can be obtained without requiring the subject to attentively listen to the speech stimulus (Vanthornhout et al. 2019), which is useful in designing a clinical test protocol, in particular for small children. Importantly, the modeling of the brain-speech interaction could be improved by considering the effect of focal attention on cortical tracking of the speech envelope, objectively quantified using a measure of EEG entropy (Lesenfants and Francart 2019). While the objective measure of speech intelligibility has not yet been measured in children, there is evidence that neural tracking of the speech envelope is present in young children (Ríos-López et al. 2020; Vander Ghinst et al. 2019). The objective measure for speech intelligibility can be measured across the adult lifespan: while neural tracking of speech envelope increases with age (Presacco et al. 2016), young, middle-aged and older adults also show increases in envelope tracking with increasing speech intelligibility (Decruy et al. 2019a). Furthermore, age-matched hearing-impaired adults show an additional increase in neural envelope tracking compared to their normal-hearing peers. Nevertheless, envelope tracking also increases as a function of speech intelligibility for adults with a disabling hearing loss (Decruy et al. 2020). Additionally, there are indications that neural envelope tracking in the delta band (0.5 to 4 Hz) is not confounded by listening effort (Decruy et al. 2019b).

## 3 EEG Recordings in Cochlear Implant Users

To establish a closed-loop CI system that interfaces with the brain, neural activity needs to be recorded while the implant is stimulating the auditory nerve. EEG is seen as a promising technology for this purpose, as it is relatively cheap and portable. However, the electrical stimulation by the implant causes artifacts that may obscure the recorded neural responses.

Many evoked response paradigms make use of short click or burst stimuli which are repeated multiple times. Because the neural response occurs with a delay after the stimulus, the contamination by the artifact is small or can be reduced with relatively simple methods. For instance, electrical pulses of alternating polarities can be used over different trials, causing the resulting artifacts to partially cancel out when averaging over trials. Other methods based on modelling the artifact with a template or an exponential fit and subsequently subtracting it are also effective in reducing the artifact (Brown and Abbas 1990; Hofmann and Wouters 2010; McLaughlin et al. 2012, 2013).

In paradigms with continuously ongoing stimulation, such as Auditory Steady State Responses (ASSRs), there is no longer a temporal difference between the instantaneous artifact and the delayed neural response that can be easily exploited. The current state of the art methods for artifact removal are based on limiting the pulse rate of the continuous stimulation (e.g. 500 pulses per second) (Deprez et al. 2014; Gransier et al. 2016; Hofmann and Wouters 2010, 2012). As this prevents the artifacts from overlapping, EEG samples between subsequent artifact pulses can be used for detecting artifact-free responses. In such paradigms, there are still multiple trials which can be averaged to enhance the response SNR.

Translating the objective measure of speech intelligibility to CI users is challenging. Presenting intelligible speech to CI users requires continuous stimulation at high pulse rates. Furthermore, the EEG experiments are usually carried out as a single-trial paradigm: the natural speech stimulus (e.g. a story or audiobook) is only presented once without repetitions. We developed a method to eliminate the continuously ongoing and temporally overlapping artifacts based on periodic blanking of the stimulus, without affecting the intelligibility of the stimulus (Somers et al. 2019). This enabled the measurement of neural envelope tracking in CI users. Further experiments using this technique showed that neural envelope tracking also correlates with speech intelligibility in CI users (Verschueren et al. 2019a), demonstrating the potential of neural envelope tracking as an objective measure of speech intelligibility in CI users.

## 4 Intracochlear Neural Recordings

Another essential component of a closed-loop CI is the functionality to continuously record brain responses to speech in an unobtrusive way. An elegant solution is to use the electrodes of the cochlear implant itself. This removes the need for scalp electrodes and enables chronic monitoring of the user outside of the clinic. While current commercial CIs already have basic telemetry capabilities, they are limited to short recording windows, which can only capture very peripheral auditory measures related to neural survival in the cochlea and auditory nerve (Tejani et al. 2019). Objective measures that capture auditory processing on the cortical level require longer recording windows (McLaughlin et al. 2012), or even continuous record-ings such as our objective measure of speech intelligibility using running speech (Vanthornhout et al. 2018). Some studies have demonstrated the recording of neural responses evoked by auditory stimulation from implanted, non-cochlear electrodes, for instance with a few surgically placed epidural electrodes (Haumann et al. 2019) or an intracranial grid (Nourski et al. 2013). However, these electrodes were only placed temporarily and were highly invasive.

To develop an EEG recording technique using implanted CI electrodes, we recruited CI users with an experimental device containing a percutaneous connector, providing direct access to the implanted electrodes (Somers et al. 2020). They are enrolled in a program Cochlear Ltd., a leading CI manufacturer. This allowed us to perform continuous EEG recordings from the implanted electrodes while simulta-neously stimulating the other electrodes on the implant. Additionally, we applied conventional scalp electrodes on the head to compare the recordings with the clin-ical montages (Fig. 4a). We characterized the intra-cochlear EEG and how it is influenced by acquisition parameters using electrically evoked auditory brainstem responses and cortical evoked potentials, and varied parameters such as the stimula-tion/recording electrode pairs. Furthermore, we investigated removal of the stimulus artifact, which is exceptionally large when recording in the cochlea. The targeted neural responses were successfully recovered from the continuous EEG recordings. Figure 4b shows some long-latency cortical response complexes measured using the percutaneous plug setup in various recording configurations. These results allow us

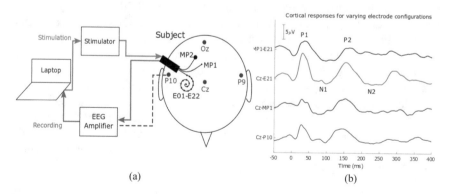

(a)                                                                                          (b)

**Fig. 4  a** Schematic overview of percutaneous recording setup and electrode locations on a subject. **b** Cortical evoked responses measured from implanted electrodes of CI. E21 is an intracochlear electrode, MP1 is an implanted extracochlear reference electrode, Cz and P10 are scalp electrodes. The stimulus occurred at 0 ms. The P1, N1, P2 and N2 response peaks are indicated

to make informed decisions regarding the requirements for embedding a BCI system in cochlear implants, such as the optimal recording electrode configurations and amplifier specifications.

## 5 Future Directions—Closed-Loop Cochlear Implants

Closing the loop in cochlear implants by integrating a BCI into the system requires a method to adjust the electrical stimulation based on the neural feedback: if the user's speech intelligibility is measured to be poor from the objective measures, how should stimulation be changed to improve it? Not much research has been done yet for cochlear implants; however, the concept of neuro-steered hearing aids has been explored. Hearing aids perform a frequency-specific amplification of the sound and stimulate acoustically to alleviate hearing loss. EEG responses to continuous speech have mainly been used in this context for auditory attention detection: to which of multiple speakers is a (hearing-impaired) person trying to attend? It has been shown that the brain processes different speech streams as separate auditory objects (Shinn-Cunningham 2008) and that attended speech is represented stronger when decoded (O'Sullivan et al. 2015). These responses have been used to steer noise suppression and gain control algorithms for application in neuro-steered hearing aids (Aroudi and Doclo 2020; Das et al. 2018; Geirnaert et al. 2020; Van Eyndhoven et al. 2017). These methods can also be translated to neuro-steered CIs.

When it comes to the actual fitting of cochlear implants, i.e. adapting the conversion of sound to electrical stimulation rather than improving SNR at the microphone input with the methods described above, paradigms based on evoked potentials have been proposed (Finke et al. 2017; Visram et al. 2015). However, these evoked potentials make use of artificial non-speech stimuli and therefore don't predict speech intelligibility well: they are generally suited for detection of thresholds and loudness growth functions. Measures reflecting speech intelligibility will likely serve as a better (additional) input to automated fitting algorithms. There is a large parameter space of subject-dependent settings that affect CI outcomes such as speech intelligibility, e.g. pulse rate, pulse width and threshold/comfortable level profiles. Due to restrictions, many of these are not assessed during a behavioral fitting procedure in the clinic and are kept at default values. Genetic algorithms have been proposed in the literature to explore the large parameter space and determine a good fitting (Durant et al. 2004; Wakefield et al. 2005). The main challenge in these studies was the collection of enough behavioral subject responses and their variability. This issue can be overcome in a closed-loop CI system that continuously measures neural responses in everyday life.

# 6  Conclusion

The research presented here gives an overview of recent scientific progress towards closed-loop cochlear implants. In such systems, a brain-computer interface is embedded into the cochlear implant to monitor neural responses, process them into relevant measures of hearing outcomes, and automatically adapt the stimulation settings. For unobtrusive, continuous monitoring of neural responses, the implanted electrodes of the CI itself can be used. Together with audio signals recorded by the microphone, the neural responses can be processed into an objective measure of speech intelligibility, as speech intelligibility is still the most important hearing outcome to optimize for CI users. In the future, closed-loop cochlear implants have the potential to provide a better and more comfortable hearing experience for severely hearing-impaired persons.

# References

S.J. Aiken, T.W. Picton, Human cortical responses to the speech envelope. Ear Hear. (2008). https://doi.org/10.1097/AUD.0b013e31816453dc

A. Aroudi, S. Doclo, Cognitive-driven binaural beamforming using eeg-based auditory attention decoding. IEEE/ACM Trans. Audio Speech Lang. Process. (2020). https://doi.org/10.1109/taslp.2020.2969779

J. Brown, P.J. Abbas, Electrically evoked whole-nerve action potentials: Data from human cochlear implant users. J. Acoust. Soc. Am. **88**(September 1990), 1385–1391 (1990)

N. Das, A. Bertrand, T. Francart, EEG-based auditory attention detection: Boundary conditions for background noise and speaker positions. J. Neural Eng. (2018). https://doi.org/10.1088/1741-2552/aae0a6

N. Das, J. Vanthornhout, T. Francart, A. Bertrand, Stimulus-aware spatial filtering for single-trial neural response and temporal response function estimation in high-density EEG with applications in auditory research. NeuroImage (2020). https://doi.org/10.1016/j.neuroimage.2019.116211

L. Decruy, N. Das, E. Verschueren, T. Francart, The self-assessed békesy procedure: validation of a method to measure intelligibility of connected discourse. Trends Hear. (2018). https://doi.org/10.1177/2331216518802702

L. Decruy, D. Lesenfants, J. Vanthornhout, T. Francart, Top-down modulation of neural envelope tracking: the interplay with behavioral, self-report and neural measures of listening effort. BioRxiv (2019a). https://doi.org/10.1101/815365

L. Decruy, J. Vanthornhout, T. Francart, Evidence for enhanced neural tracking of the speech envelope underlying age-related speech-in-noise difficulties. J. Neurophysiol. (2019b). https://doi.org/10.1152/jn.00687.2018

L. Decruy, J. Vanthornhout, T. Francart, Hearing impairment is associated with enhanced neural tracking of the speech envelope. BioRxiv (2020). https://doi.org/10.1101/815530

H. Deprez, M. Hofmann, A. Van Wieringen, J. Wouters, M. Moonen, Cochlear implant artifact rejection in electrically evoked auditory steady state responses. 22nd European Signal Processing Conference (EUSIPCO), 1995–1999 (2014). https://doi.org/10.1017/CBO9781107415324.004

G.M. Di Liberto, J.A. O'Sullivan, E.C. Lalor, Low-frequency cortical entrainment to speech reflects phoneme-level processing. Curr. Biol. **25**(19), 2457–2465 (2015). https://doi.org/10.1016/j.cub.2015.08.030

G.M. Di Liberto, E.C. Lalor, Isolating neural indices of continuous speech processing at the phonetic level. Adv. Exp. Med. Biol. (2016). https://doi.org/10.1007/978-3-319-25474-6_35

N. Ding, J.Z. Simon, Neural coding of continuous speech in auditory cortex during monaural and dichotic listening. J. Neurophysiol. **107**(1), 78–89 (2012). https://doi.org/10.1152/jn.00297.2011

N. Ding, J.Z. Simon, Adaptive temporal encoding leads to a background-insensitive cortical representation of speech. J. Neurosci. **33**(13), 5728–5735 (2013). https://doi.org/10.1523/JNEURO SCI.5297-12.2013

M.F. Dorman, B.S.B.S. Wilson, The design and function of cochlear implants. Am. Sci. **92**(6), 436–445 (2004). https://doi.org/10.1511/2004.49.942

E.A. Durant, G.H. Wakefield, D.J. Van Tasell, M.E. Rickert, Efficient perceptual tuning of hearing aids with genetic algorithms. IEEE Trans. Speech Audio Process. (2004). https://doi.org/10.1109/TSA.2003.822640

M. Finke, M. Billinger, A. Büchner, Toward automated cochlear implant fitting procedures based on event-related potentials. Ear Hear. **38**(2), e118–e127 (2017). https://doi.org/10.1097/AUD.000 0000000000377

T. Francart, A. Van Wieringen, J. Wouters, Comparison of fluctuating maskers for speech recognition tests. Int. J. Audiol. (2010). https://doi.org/10.3109/14992027.2010.505582

S. Geirnaert, T. Francart, A. Bertrand, An interpretable performance metric for auditory attention decoding algorithms in a context of neuro-steered gain control. IEEE Trans. Neural Syst. Rehabil. Eng. (2020). https://doi.org/10.1109/TNSRE.2019.2952724

R. Gransier, H. Deprez, M. Hofmann, M. Moonen, A.Van Wieringen, J. Wouters, Auditory steady-state responses in cochlear implant users: effect of modulation frequency and stimulation artifacts. Hear. Res. **335**, 1–43 (2016). https://doi.org/10.1016/j.heares.2016.03.006

S. Haumann, G. Bauernfeind, M.J. Teschner, I. Schierholz, M.G. Bleichner, A. Büchner, T. Lenarz, Epidural recordings in cochlear implant users. J. Neural Eng. (2019). https://doi.org/10.1088/1741-2552/ab1e80

M. Hofmann, J. Wouters, Electrically evoked auditory steady state responses in cochlear implant users. J. Assoc. Res. Otolaryngol.: JARO **11**(2) (2010). https://doi.org/10.1007/s10162-009-0201-z

M. Hofmann, J. Wouters, Improved electrically evoked auditory steady-state response thresholds in humans. J. Assoc. Res. Otolaryngol. **13**(4), 573–589 (2012)

D. Lesenfants, T. Francart, The interplay of top-down focal attention and the cortical tracking of speech. BioRxiv (2019). https://doi.org/10.1101/813147

D. Lesenfants, J. Vanthornhout, E. Verschueren, L. Decruy, T. Francart, Predicting individual speech intelligibility from the cortical tracking of acoustic- and phonetic-level speech representations. Hear. Res. (2019). https://doi.org/10.1016/j.heares.2019.05.006

M. McLaughlin, A. Lopez Valdes, R.B. Reilly, F.-G. Zeng, Cochlear implant artifact attenuation in late auditory evoked potentials: a single channel approach. Hear. Res. **302**, 84–95 (2013). https://doi.org/10.1016/j.heares.2013.05.006

M. McLaughlin, T. Lu, A. Dimitrijevic, F.G. Zeng, Towards a closed-loop cochlear implant system: Application of embedded monitoring of peripheral and central neural activity. IEEE Trans. Neural Syst. Rehabil. Eng. **20**(4), 443–454 (2012). https://doi.org/10.1109/TNSRE.2012.2186982

K.V. Nourski, C.P. Etler, J.F. Brugge, H. Oya, H. Kawasaki, R.A. Reale, P.J. Abbas, C.J. Brown, M.A. Howard, Direct recordings from the auditory cortex in a cochlear implant user. JARO: J. Assoc. Res. Otolaryngol. (2013). https://doi.org/10.1007/s10162-013-0382-3

J.A. O'Sullivan, A.J. Power, N. Mesgarani, S. Rajaram, J.J. Foxe, B.G. Shinn-Cunningham, M. Slaney, S.A. Shamma, E.C. Lalor, Attentional selection in a Cocktail Party Environment can be decoded from single-trial EEG. Cereb. Cortex **25**(7), 1697–1706 (2015). https://doi.org/10.1093/cercor/bht355

J.E. Peelle, M.H. Davis, Neural oscillations carry speech rhythm through to comprehension. Front. Psychol. (2012). https://doi.org/10.3389/fpsyg.2012.00320

A. Presacco, J.Z. Simon, S. Anderson, Evidence of degraded representation of speech in noise, in the aging midbrain and cortex. J. Neurophysiol. (2016). https://doi.org/10.1152/jn.00372.2016

P. Ríos-López, N. Molinaro, M. Bourguignon, M. Lallier, Development of neural oscillatory activity in response to speech in children from four to six years old. Dev. Sci. (2020). https://doi.org/10.1111/desc.12947

R.V. Shannon, F.G. Zeng, V. Kamath, J. Wygonski, M. Ekelid, Speech recognition with primarily temporal cues. Science (1995). https://doi.org/10.1126/science.270.5234.303

B.G. Shinn-Cunningham, Object-based auditory and visual attention. Trends Cogn. Sci. (2008). https://doi.org/10.1016/j.tics.2008.02.003

B. Somers, T. Francart, A. Bertrand, A generic EEG artifact removal algorithm based on the multi-channel Wiener filter. J. Neural Eng. (2018). https://doi.org/10.1088/1741-2552/aaac92

B. Somers, C.J. Long, Francart, EEG-based diagnostics of the auditory system using cochlear implant electrodes as sensors. BioRxiv (2020)

B. Somers, E. Verschueren, T. Francart, Neural tracking of the speech envelope in cochlear implant users. J. Neural Eng. (2019). https://doi.org/10.1088/1741-2552/aae6b9

V.D. Tejani, P.J. Abbas, C.J. Brown, J. Woo, An improved method of obtaining electrocochleography recordings from Nucleus Hybrid cochlear implant users. Hear. Res. (2019). https://doi.org/10.1016/j.heares.2019.01.002

S. Van Eyndhoven, T. Francart, A. Bertrand, EEG-informed attended speaker extraction from recorded speech mixtures with application in neuro-steered hearing prostheses. IEEE Trans. Biomed. Eng. (2017). https://doi.org/10.1109/TBME.2016.2587382

M. Vander Ghinst, M. Bourguignon, M. Niesen, V. Wens, S. Hassid, G. Choufani, V. Jousmäki, R. Hari, S. Goldman, X. De Tiège, Cortical tracking of speech-in-noise develops from childhood to adulthood. J. Neurosci. (2019). https://doi.org/10.1523/JNEUROSCI.1732-18.2019

J. Vanthornhout, L. Decruy, T. Francart, Effect of task and attention on neural tracking of speech. Front. Neurosci. (2019). https://doi.org/10.3389/fnins.2019.00977

J. Vanthornhout, L. Decruy, J. Wouters, J.Z. Simon, T. Francart, Speech intelligibility predicted from neural entrainment of the speech envelope. JARO: J Assoc. Res. Otolaryngol. (2018). https://doi.org/10.1007/s10162-018-0654-z

E. Verschueren, B. Somers, T. Francart, Neural envelope tracking as a measure of speech understanding in cochlear implant users. Hear. Res. (2019a). https://doi.org/10.1016/j.heares.2018.12.004

E. Verschueren, J. Vanthornhout, T. Francart, The effect of stimulus choice on an EEG-based objective measure of speech intelligibility. BioRxiv (2019b). https://doi.org/10.1101/421727

A.S. Visram, H. Innes-Brown, W. El-Deredy, C.M. McKay, Cortical auditory evoked potentials as an objective measure of behavioral thresholds in cochlear implant users. Hear. Res. (2015). https://doi.org/10.1016/j.heares.2015.04.012

G.H. Wakefield, C. Van Den Honert, W. Parkinson, S. Lineaweaver, Genetic algorithms for adaptive psychophysical procedures: Recipient-directed design of speech-processor MAPs. Ear Hear. (2005). https://doi.org/10.1097/00003446-200508001-00008

J. Wouters, H.J. Mcdermott, T. Francart, Sound coding in cochlear implants: from electric pulses to hearing. IEEE Signal Process. Mag. **32**(2), 67–80 (2015). https://doi.org/10.1109/MSP.2014.2371671

F.G. Zeng, S. Rebscher, W. Harrison, X. Sun, H. Feng, Cochlear implants: System design, integration, and evaluation. IEEE Rev. Biomed. Eng. (2008). https://doi.org/10.1109/RBME.2008.2008250

# Final Results of Multi-center Randomized Controlled Trials of BCI-Controlled Hand Exoskeleton Complex Assisting Post-stroke Motor Function Recovery

**Alexander Frolov, Elena Biryukova, Pavel Bobrov, Dmirty Bobrov, Alexander Lekin, Olesya Mokienko, Roman Lyukmanov, Sergey Kotov, Anna Kondur, Galina Ivanova, and Yulia Bushkova**

**Abstract** The project is aimed at investigating efficacy of a BCI-controlled palm exoskeleton as a tool for motor function recovery in post-stroke patients. The idea of using the system is grounded on vast amount of data supported by physiologic literature and our own findings in healthy subjects, suggesting that kinesthetic motor imagery (MI) requires activation of the brain areas involved in motion planning, execution and control. Thus, the common idea of using a MI-based BCI for neurorehabilitation is to reinforce motor imagery of intention to move with visual, proprioceptive and\or tactile feedback. Results of a four-year multi-center randomized controlled study of post-stroke motor rehabilitation procedure with BCI-controlled hand exoskeleton complex are presented. The study has the largest number of participants so far. Statistical analysis of different clinical scales used to assess motor function recovery show that incorporating the BCI+exoskeleton procedure into rehabilitation significantly improves its outcome. The analysis also revealed non-monotonical dependency of motor function recovery rate on initial motor and sensory function status, as well as on age, and BCI control accuracy. Hopefully, the reported data combined with the results obtained by other groups in the world, would provide solid evidence supporting inclusion of the BCI-based systems into rehabilitation practice.

A. Frolov · E. Biryukova · P. Bobrov (✉) · D. Bobrov · A. Lekin · G. Ivanova · Y. Bushkova
Pirogov Russian National Research Medical University, Moscow, Russia
e-mail: bobrov.pavel@ihna.ru

A. Frolov · E. Biryukova · P. Bobrov
Institute of Higher Nervous Activity, Moscow, Russia

O. Mokienko · R. Lyukmanov
Research Centre of Neurology, Moscow, Russia

S. Kotov · A. Kondur
Vladimirsky Moscow Regional Research Clinical Institute, Moscow, Russia

© The Author(s), under exclusive license to Springer Nature Switzerland AG 2021
C. Guger et al. (eds.), *Brain-Computer Interface Research*,
SpringerBriefs in Electrical and Computer Engineering,
https://doi.org/10.1007/978-3-030-60460-8_6

**Keywords** Brain-computer interface · Post-stroke rehabilitation · Clinical trials ·
Hand exoskeleton · Motor imagery

# 1 Introduction

According to the World Health Organization, stroke incidences varies from 50 to
500 per 100,000 population depending on the world region, and the rate of stroke
mortalities amounts about 50% of stroke incidence (Thrift et al. 2014). Consequently,
tens of millions of people in the world suffer from the effects of stroke and mainly
from motor disorders (Paolucci et al. 2000). Thus, the search of approaches to recovering the motor functions in post-stroke patients is one of the most important tasks in
neurorehabilitation (Langhorne et al. 2009; Pollock et al. 2015). However, none of
the existing motor rehabilitation methods is assigned the highest evidence level and
recommendation grade for motor recovery. Levels 2–3 of evidence for post-stoke
upper extremity function recovery are demonstrated by a virtual reality technology,
robotic methods, constraint-induced movement therapy and "mental trainings", in
particular motor imagery (Langhorne et al. 2009; Pollock et al. 2015). It should
be emphasized that techniques involving active motor paradigms, such as robotic
methods and constraint therapy, are applicable only in mild or moderate paresis.
In the case of either severe paresis or plegia, only motor imagery seems to be a
plausible technique to stimulate the mechanisms of brain plasticity directed to the
motor recovery. As shown in many works (Jeannerod 1994, 2001; Solodkin et al.
2004), motor imagery follows the same principles as the motor execution and, therefore, is likely to stimulate the brain plasticity by the same mechanisms as the actual
execution of movements (Frolov et al. 2016a; Mokienko et al. 2013). Monitoring of
the motor imagery can be done with the help of a brain-computer interface (BCI),
which records the EEG signals of the brain resulting from the motor imagery into the
controlling commands for an external device. The command execution provides a
patient with biofeedback, allowing the patient to concentrate on performing the motor
imagery task. The BCI technology seems especially efficient if it is combined with
an exoskeleton or a manipulator as the external devices controlled via the BCI. The
patient receives not only visual feedback, but also haptic and kinesthetic feedback
that is contingent upon the imagination of a specific movement.

Several BCI studies involving this type of haptic and kinesthetic feedback have
demonstrated improvements in clinical parameters of post-stroke motor recovery
(Ang et al. 2011, 2015; Ramos-Murguialday et al. 2013; Ono et al. 2014; Frolov
et al. 2016b, 2017). The number of subjects with post-stroke upper extremity paresis
included in these studies was, however, relatively low (from 12 in [Ono et al. 2014]
to 55 in [Frolov et al. 2017]) patients. In the present paper, we report the study
investigating many more patients. Increasing the number of investigated patients,
first, allowed for more reliable estimation of the recovery efficiency, and second,
helped estimation of its dependence on patient features, such as age, gender, severity
of stroke, duration of post-stroke period, efficiency of BCI control, because it was

possible to divide patients on the large representative groups according to these features providing more reliable statistics. The study was blind, randomized and controlled. It was performed for four years in four medical centers. These medical centers were selected, first, because of the presence of a neurorehabilitation department or motor rehabilitation service and, second, the availability of post-stroke patients with various residual periods and hemiparesis of different severity. The preliminary results of the study were published in (Frolov et al. 2016b, 2017). Besides the BCI and Control groups described in these papers, we added here a Comparison group in which patients obtained only routine medical treatment.

## 2  Study Design

The study was approved by the Ethics Committee of the Research Center of Neurology: #12/14 of 10 December 2014. All patients provided a signed informed consent for participation in the study. The study protocol was registered in the system clinicaltrials.gov: NCT02325947.

The study had the following inclusion criteria: male or female patients which underwent inpatient treatment at the study centers, aged from 18 to 80 years, with subacute (1–6 months) or chronic (more than 6 months from onset) stroke; hand paresis, mild to severe plegia, according to the Medical Research Council Sum Score scale (MRCSS); a single focus of ischemic or hemorrhagic stroke with a supratentorial localization (according to MRI or CT data); and a signed informed consent. Such a heterogeneous group was chosen in order to find a target group of patients for which the BCI procedure is the most efficient. The exclusion criteria were as follows: left-handedness according to the Edinburgh Handedness Inventory; severe cognitive impairment (<10 points according to the Montreal Cognitive Assessment Scale); sensory aphasia; severe motor aphasia; severe vision impairment preventing execution of visual instructions shown on the computer screen; muscle spasticity in the upper extremity more than 3 points according to the Modified Ashworth Scale (MAS, 1–5 points).

The withdrawal criteria were as follows: patient refusal to continue participating in the study; development of an acute disease or decompensation of a chronic disease with the risk of a potential impact on the study results (repeated stroke, acute myocardial infarction, non-compensated diabetes, etc.); prescription of systemic muscle relaxants or changing their dose after inclusion in the study; injection of botulinum toxin agents in muscles of the paretic upper extremity after inclusion of the patient to the study.

The examination data of patients who signed the informed consent and met the inclusion criteria were uploaded to an automated system for clinical research information support (Imagery Soft, Russia). The system assigned an identification number (ID) to each study participant. The IDs were randomized so that patients were assigned either to the BCI group with probability 2/3 or to the Control group with probability 1/3. The Comparison group was formed later using to the same inclusion criteria as the BCI and Control groups. The number of patients in the Comparison group was the same as in the Control group. The reduced number of patients in the Control and Comparison groups compared with the BCI group is the result of tradeoff between the intention to increase the number of patients undergoing the intensive motor imagery training and maintaining sufficient statistical power of the study.

The patients of the BCI and Control groups underwent the procedures in three clinics. The procedures for the BCI and Control groups differed in the way the exoskeletons were controlled. In the BCI group, the exoskeleton movements were controlled via the BCI system. In the Control group, exoskeleton-driven hand movements were not linked to the patients' brain activity but followed a repetitive scheme. The patients in each group performed 10 daily sessions. Each session lasted for 30–45 min. The sessions were conducted every day with breaks on weekends and holidays (up to 3 consecutive days) so the total hospitalization duration was two weeks. Patients in both groups were also provided with standard physical therapy: instructor-supervised kinesiotherapy, medical massage, and passive neuromuscular electrical stimulation in accordance with Russian treatment protocols and standards. The patients of the Comparison group were only provided with standard treatment in the fourth clinic.

## 3 The BCI Group Protocol

The patient sat on a chair in front of a computer monitor with the arms on the armrests of the chair in a comfortable position. Two exoskeletons were attached to the patient's hands. The patient was instructed by cues on the monitor to either sit relaxed or perform imagery of slowly expanding his\her left or right hand. The three classes of mental activity were discriminated by the BCI classifier, since the protocol required the patient to perform a lateralized motor imagery of both hands rather than to imagine movement of the impaired hand only. This setup also prevented the BCI from triggering the exoskeleton movement based on recognizing a mental state different from motor imagery, e.g. level of high concentration opposed to the relaxed state.

At earlier stages of the study exoskeletons with pneumatic actuators (Neurobotics, Russia) were used, while exoskeletons with electromotor actuators (Android Technics, Russia) were used for most the later sessions. The exoskeletons of both types were completely equivalent functionally, however, the last ones were more reliable.

EEG signals were recorded with 30 electrodes placed according to the International 10–20 system (NVX52, Medical Computer Systems, Zelenograd, Russia). EEG corresponding to the different mental tasks was classified using the Bayesian classifier (Frolov et al. 2011). The information on classification accuracy is given by a confusion matrix (Frolov et al. 2011). Diagonal elements of the matrix allow to estimate average probability of correct classification, which is a measure of the BCI control accuracy. It depends on both the classifier performance and the participant's ability to perform motor imagery. The chance level for correct classifying three different mental state is 33%. Our previous papers (Frolov et al. 2016b, 2017) provide more details of BCI protocol.

## 4   The Control Group Protocol

In experiments with the Control group, we used the same arrangements as in the BCI sessions, including putting the EEG cap on the patient's head and fixing the hands to the exoskeletons, but hand exoskeletons movements were not dependent on motor imagery-related EEG modulations. The patients were sitting relaxed while watching for changes in the instruction on the monitor. The cues corresponding to right and left hand were shown randomly. If the cue corresponded to movement of one of the hands, the exoskeleton opened and closed the hand periodically independently of EEG activity, which was recorded for the Control group as well as for the BCI group for off-line analysis.

## 5   Clinical Assessment and Statistical Analysis

The patients in the BCI and Control groups were assessed for movements and strength in the upper limb before and after the total training course. The patients in the Comparison group were assessed at the beginning and at the end of two weeks hospitalization; that is, for all groups the time between two assessments was the same. Motor recovery was assessed using the Fugl-Meyer Motor Assessment (FMMA) for upper extremity (range, 0–126) and Action Research Arm Test (ARAT; range, 0–57). Additionally, the changes across different FMMA domains were analyzed. The spasticity severity was assessed using Modified Ashworth Scale (MAS) and hand paresis by MRCSS. We also estimated the percentage of patients with clinically significant improvement exceeding minimal clinically important difference (MCID) in each study group. As recommended in the literature, MCID was chosen separately for subacute and chronic stroke. MCID for the ARAT scale is accepted to be a 12-point increase for dominant and 17-point increase for non-dominant hand in case of subacute stroke, and 6-point increase in case of chronic stroke. MCID for the FMMA motor domain is accepted to be a 10-point increase in case of subacute stroke and a 5-point increase in case of chronic stroke.

Statistical analysis was performed using Wilcoxon test and regression analysis using MATLAB. We also used ANOVA analysis to estimate preliminary the effects of different factors, although the data were not normally distributed. Utilizing ANOVA was justified by the fact that, for each considered factor, the corresponding groups consisted of at least 10 patients. All results revealed by ANOVA were checked by the Wilcoxon test. In order to compare the binary data, i.e. the data which could be represented by two values (0 and 1), such as gender, stroke lateralization etc., we used a test based on an assumption that the data of the two compared groups come from Bernoulli distribution with parameters $p_1$ and $p_2$ respectively. The data are presented as a median and 25 and 75% quartiles. Statistically significant differences were considered at $p < 0.05$. They are marked in bold red in the tables.

# 6 Patient Group Characteristics

The study inclusion criteria were met by 171 patients of 841 screened in total. The BCI group consisted of 92 patients, the Control group consisted of 41 patients, and the Comparison group consisted of 38 patients. Eleven patients of the BCI group and four patients of the Control group were withdrawn from the study after 1 or 2 procedures due to either their refusal or clinical requirements. Their results were excluded from the analysis.

The patients' demographics and baseline characteristics are presented in Table 1.

**Table 1** Demographics and baseline characteristics of subjects by study group

| Characteristic | BCI group (n = 81) | Control group (n = 37) | Comparison group (n = 38) | Significance (p-value)[a] | | |
|---|---|---|---|---|---|---|
| | | | | $p_{12}$ | $p_{13}$ | $p_{23}$ |
| Age, full years | 59 [50.5; 68] | 59 [57; 66.5] | 66.5 [55; 71] | 0.27 | **0.008** | 0.12 |
| Men, %(n) | 64.2 (52) | 70.3 (26) | 55.3 (21) | 0.50 | 0.36 | 0.18 |
| Time since stroke in months | 6.5 [3.5; 13] | 8 [2.5; 13] | 5 [3; 10] | 0.81 | 0.26 | 0.56 |
| Patients with subacute stroke (1–6 months), %(n) | 50.6 (41) | 59.4 (22) | 47.3 (18) | 0.37 | 0.95 | 0.42 |
| Patients with the lesion in the left hemisphere, %(n) | 48.2 (39) | 59.5 (22) | 50 (19) | 0.26 | 0.85 | 0.42 |
| Ischemic stroke, %(n) | 71.6 (58) | 89 (33) | 97.4 (37) | **0.015** | **<10⁻⁴** | 0.16 |
| Subcortical lesions, %(n) | 61.7 (50) | 48.6 [18] | 73 (28) | 0.19 | 0.19 | **0.03** |
| Initial ARAT score | 4 [0; 34.5] | 27.5 [1; 40] | 15 [2.5; 33] | 0.18 | 0.26 | 0.53 |
| Initial FMMA score | 76 [60; 95] | 93 [62; 66.5] | 79.5 [67; 98.5] | 0.08 | 0.39 | 0.40 |
| Spasticity (MAS) | 2 [1; 3] | 1.5 [1; 2] | 1 [1; 2] | **0.04** | **0.02** | 0.95 |
| MRCSS | 2 [1; 3] | 3 [1; 4] | 3 [2; 3] | 0.20 | **0.01** | 0.75 |

[a] According to Bernoulli test for the binary and Wilcoxon test for the non-binary data

For binary data, only one category is presented. There were only a few patients with lesion localization that was considered as cortical by the clinicians, 4 in the BCI group, 7 in the Control group and none in the Comparison group. They were included into the group of patients with cortico-subcortical lesions. Therefore, the localization factor became binary (subcortical or cortico-subcortical).

The groups do not differ significantly with respect to most of the factors. Exceptions were that the patients from the Comparison group were significantly older than the patients from the other groups, there were significantly more hemorrhagic stroke cases in the BCI group compared to the other groups, and the average initial spasticity was higher in the BCI group.

General population analysis revealed that there were significantly more males (63%) than females ($p < 10^{-4}$), more ischemic than hemorrhagic stroke cases (82%, $p < 10^{-5}$), and number of subcortical lesioned patients (61%) exceeded the number of cortico-subcortical lesioned patients ($p = 3 \cdot 10^{-3}$).

## 7  Rehabilitation Outcome

Table 2 contains differences (gains) in the scales used to assess the patients' motor and sensory functions before and after intervention. The table contains both quartiles and p-values for pairwise group comparison. The $p$-values were obtained using both ANOVA statistics (left columns) and Wilcoxon test (right columns). The significance of MCID percentage was tested using the Bernoulli test. FMMA stands for total FMMA score, FMMA-M stands for score in the motor function domain (Domain 1), FMMA-D stands for score in the distal joint subdomains (VII and VIII), FMMA-P stands for score in the proximal joint subdomains (I–VI, IX), FMMA-S stands for sensory function (Domain 2), FMMA-JR stands for score in the range of motion domain (Domain 4), FMMA-JP stands for score in the joint pain domain (Domain 5), and ARAT stands for total ARAT score.

All scales indicate significantly higher increase of motor function scores in the BCI group compared to other groups, except for comparison to the Control group for the distal joint FMMA subdomains. The differences were significant both according to the ANOVA and Wilcoxon tests. Notably, there was a higher percentage of the patients who achieved MCID in the BCI group, which was two times higher than the other groups according to FMMA and more than seven times higher according to ARAT.

There were no significant differences between the BCI and other groups in the FMMA sensory, pain, and range of motion domains. There were also no significant differences between the Control and Comparison groups.

**Table 2** Clinical outcome for all study groups

| Score | BCI group | Control group | Comparison group | Significance | p12 | p13 | p23 | | Possible range |
|---|---|---|---|---|---|---|---|---|---|
| FMMA | 6 [2; 9] | 3 [1; 6] | 3 [0; 6.5] | **0.02** | **0.003** | **0.003** | 0.80 | 0.70 | 0–126 |
| FMMA-M | 4 [1.5; 7] | 2 [0; 3] | 1 [0; 3] | **0.006** | **0.003** | **$<10^{-5}$** | 0.36 | 0.21 | 0–66 |
| FMMA-D | 2 [0; 4] | 1 [0; 2] | 0 [0; 1] | 0.16 | 0.09 | **$10^{-3}$** | 0.34 | 0.06 | 0–24 |
| FMMA-P | 2 [1; 4] | 0 [0; 1.5] | 1 [0; 2] | **0.005** | **$<10^{-3}$** | **0.05** | 0.76 | 0.14 | 0–42 |
| FMMA-S | 0 [0; 1] | 0 [0; 0] | 0 [0; 0] | 0.76 | 0.23 | 0.99 | 0.88 | 0.67 | 0–12 |
| FMMA-JR | 0 [0; 2] | 0 [0; 1] | 0 [0; 2] | 0.73 | 0.23 | 0.99 | 0.78 | 0.24 | 0–24 |
| FMMA-JP | 0 [0; 1] | 0 [0; 1] | 0 [0; 1] | 0.97 | 0.88 | 0.78 | 0.93 | 0.96 | 0–24 |
| ARAT | 2 [0; 5] | 1 [0; 2] | 0 [0; 2] | **0.02** | **0.05** | **0.01** | 0.99 | 0.91 | 0–57 |
| FMMA MCID | 45% | 16% | 11% | **$<10^{-2}$** | **$<10^{-3}$** | **$<10^{-3}$** | 0.57 | | 0–1 |
| ARAT MCID | 21% | 3% | 0% | **$<10^{-3}$** | **$<10^{-4}$** | **$<10^{-3}$** | 0.3 | | 0–1 |
| MAS | 0 [−1; 0] | 0 [0; 0] | 0 [0; 0] | **0.05** | **0.05** | **$<10^{-3}$** | 0.7 | 0.06 | 0–1 |

# 8  Factors That Affect Motor Function Recovery in the BCI Group

We investigated the possible effects of different factors on motor recovery status in the BCI group. FMMA and ARAT gains were chosen as the motor recovery indicators. The factors were age, gender, residual period, stroke type, lesion lateralization and localization, FMMA-M, FMMA-S, FMMA-JR, FMMA-JP, ARAT, MRCSS, and MAS scores prior interventions, and BCI control accuracy. The results of comparing the recovery outcomes for binary factors are shown in Table 3. The only significant difference was observed for the lateralization factor. The patients with lesions located in the left hemisphere recovered significantly better.

The possible effect of non-binary factors was tested using quadratic regression modelling. The procedure efficiency did not depend on initial MAS, FMMA-JR and FMMA-JP scores. Its dependencies on age, FMMA-M, FMMA-S, ARAT, MRCSS and BCI control accuracy for other factors are shown in Fig. 1. Except for the age, the dependence from each other factor was significant for at least one of the shown outcomes. Interestingly, these dependencies were non-monotonical, which is more evident for FMMA-M outcome. Dependence on age was at the level of tendency.

**Table 3**  Clinical outcomes in the BCI group with respect to the binary factors

| Factor | FMMA-M gain | ARAT gain | FMMA-M gain | ARAT gain | $p_{FMMA}$ | $p_{ARAT}$ |
|---|---|---|---|---|---|---|
| Gender | Female, n = 29 | | Male, n = 52 | | | |
| | 6.0 [2.0; 7.5] | 1.0 [0; 3.0] | 6.5 [4.0; 11.0] | 2.0 [0; 6.0] | 0.31 | 0.17 |
| Residual period | Subacute, n = 41 | | Chronic, n = 40 | | | |
| | 6.0 [2.0; 8.5] | 1.0 [0.0; 3.0] | 6.0 [2.0; 10.0] | 2.0 [0.0; 5.8] | 0.94 | 0.60 |
| Stroke type | Ischemic, n = 58 | | Hemorrhagic, n = 23 | | | |
| | 4.0 [1.5; 7.0] | 1.0 [0; 2.0] | 4.0 [2.0; 11] | 1.0 [0; 6.0] | 0.25 | 0.27 |
| Lesion localization | Subcortical, n = 50 | | Cortico-subcortical, n = 31 | | | |
| | 4.0 [1.0; 9.0] | 1.0 [0; 3.0] | 4.0 [2.0; 7.0] | 1.0 [0; 3.0] | 0.91 | 0.87 |
| Lesion lateralization | Left hemisphere, n = 39 | | Right hemisphere, n = 42 | | | |
| | 7.0 [4.5; 12.0] | 2.0 [1.0; 6.0] | 4 [2.0; 7.0] | 1.0 [0; 2.0] | **0.02** | **0.02** |

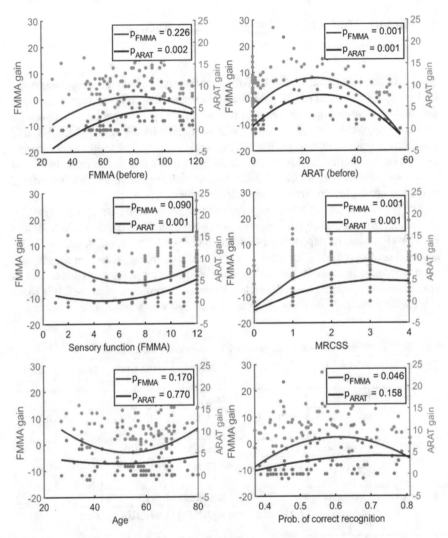

**Fig. 1** Dependence of clinical outcomes from different non-binary factors

## 9 Discussion

This work concludes four years of clinical trials of post-stroke motor rehabilitation procedures with a BCI-controlled hand exoskeleton complex. The study methodology included two scales for assessing recovery of upper extremity motor function and the recruitment of patients of BCI and Control groups from 3 clinical centers. FMMA is the more versatile and detailed scale (Ang et al. 2014; Sanford et al. 1993), while ARAT is a functional scale and evaluates different hand movements needed for daily tasks (Doussoulin et al. 2012). Thus, coincidence of results obtained by

2 scales increases their reliability. Testing of patients by specialists from different clinical centers and applying a blind study design reduced the influence of subjective factors (Sanford et al. 1993) on the assessment of clinical test performance.

Preliminary results of the study were published in (Frolov et al. 2016b, 2017) and included BCI and Control groups. The number of patients was increased from 55 to 81 in the BCI group and from 19 to 37 in the Control group, and a new Comparison group of patients receiving routine treatment was added. Thus, the results presented in this work are obtained from larger datasets than those published before (Ang et al. 2011, 2015; Ramos-Murguialday et al. 2013; Ono et al. 2014; Frolov et al. 2016b, 2017). Motor function improvement was significantly higher in the BCI group compared to the other groups for almost all the indices considered. At the same time, no significant differences in motor function recovery were observed when the Control and the Comparison groups were matched. The results suggest the BCI procedure is effective, which might result from coupling the patients' mental attempts to imagine movement with actual proprioceptive and tactile feedback.

The results in general agree with those obtained before. However, increasing the number of patients allowed us to investigate possible effect of different factors on clinical outcome of the tested procedure.

We have considered the factors of age, gender, residual period, stroke type, lesion lateralization and localization, FMMA, FMMA motor, sensory, joint motion and joint pain domains, ARAT, MRCSS, and MAS scores prior interventions and BCI control accuracy. No significant effect on clinical outcomes was observed for the factors of gender, residual period, stroke type and localization (Table 3), initial MAS, FMMA pain and joint motion range scores. Thus, the BCI procedure was effective for wide range of the patients who met the inclusion criteria. In particular, the procedure was effective in both subacute and chronic stroke patients, which is in agreement with the results of (Ono et al. 2014). The insignificance of the other factors is, to our knowledge, reported for the first time and should be further investigated in more detail. For example, exploring the effect of the lesion localization may require checking individual MRI scans to identify the lesioned areas.

The BCI procedure was significantly more effective in the patients with lesions in the left hemisphere ($p = 0.02$ for both FMMA and ARAT gains). This result also requires further investigation.

The regression analysis revealed significant dependencies of either FMMA or ARAT outcomes on initial ARAT, FMMA domain 1 and 2, and MRCSS scores, as well as on the BCI control accuracy (Fig. 1). Comparing the patients from the second and the fourth age quartiles using Wilcoxon test also revealed a significant difference ($p = 0.04$). The observed dependencies were non-monotonical and were similar for the FMMA and ARAT outcomes. The procedure was shown to be more effective for average levels of motor function impairment, which might be due to the fact that dramatic improvement of motor function is unlikely to happen after two weeks of treatment in severe lesioned patients, and the ARAT and FMMA scales are not sensitive enough to detect motor function improvement in patients with mild paresis. We have shown that motor function recovery in cases of both severe and mild paresis could be assessed with biomechanical analysis of performed and attempted movements (Dzhalagoniya et al. 2018).

The not monotonic dependence of the outcome on the BCI control accuracy may be explained as follows. The lower accuracy, which might indicate lack of effort and concentration on the mental task, results in the lack of proprioceptive feedback. The higher accuracy, e.g. typical for former sportsmen, might indicate that the task is easy. Hence, a certain level of the task difficulty as well as adequate feedback might be required to maximize the procedure efficiency.

**Acknowledgments** The study was supported by Russian Ministry of Education and Science, grant RFMEFI60519X0184.

# References

K.K. Ang, C. Guan, K.S.G. Chua, B.T. Ang, C.W.K. Kuah, C. Wang, K.S. Phua, Z.Y. Chin, H. Zhang, A large clinical study on the ability of stroke patients to use an EEG-based motor imagery brain-computer interface. Clin. EEG Neurosci. **42**(4), 253–258 (2011)

K.K. Ang, C. Guan, K.S. Phua, C. Wang, L. Zhou, K.Y. Tang, E. Joseph, J. Gopal, C.W.K. Kuah, K.S.G. Chua, Brain-computer interface-based robotic end effector system for wrist and hand rehabilitation: results of a three-armed randomized controlled trial for chronic stroke. Front. Neuroeng. **7**, 30 (2014)

K.K. Ang, K.S.G. Chua, K.S. Phua, C. Wang, Z.Y. Chin, C.W.K. Kuah, W. Low, C. Guan, A randomized controlled trial of EEG-based motor imagery brain-computer interface robotic rehabilitation for stroke. Clin. EEG Neurosci. **46**(4), 310–320 (2015)

S.A. Doussoulin, S.R. Rivas, S.V. Campos, Validation of "Action Research Arm Test" (ARAT) in Chilean patients with a paretic upper limb after a stroke. Rev. Med. Chil. **140**(1), 59–65 (2012)

I. Dzhalagoniya, E. Biryukova, Y. Bushkova, M. Kurganskaia, P. Bobrov, Biomechanical assessment of Fugl-Meyer score: the case of one post stroke patient who has undergone the rehabilitation using hand exoskeleton controlled by brain-computer interface. Int. J. Phys. Med. Rehabil. **6**(468), 2 (2018)

A. Frolov, D. Husek, P. Bobrov, Comparison of four classification methods for brain-computer interface. Neural Netw. World **21**(2), 101–115 (2011). https://doi.org/10.14311/Nnw.2011.21.007

A Frolov, D Husek, A Silchenko, J Tintera, J Rydlo, The changes in the hemodynamic activity of the brain during motor imagery training with the use of brain-computer interface. Hum. Physiol. **42** (1), 1–12 (2016a)

A Frolov, O Mokienko, LR Kh, L Chernikova, S Kotov, L Turbina, E Biryukova, A Kondur, G Ivanova, A Staritsyn, Preliminary results of a controlled study of BCI–exoskeleton technology efficacy in patients with poststroke arm paresis. Bull. Russ. State Med. Univ. (2) (2016b)

A.A. Frolov, O. Mokienko, R. Lyukmanov, E. Biryukova, S. Kotov, L. Turbina, G. Nadareyshvily, Y. Bushkova, Post-stroke rehabilitation training with a motor-imagery-based brain-computer interface (BCI)-controlled hand exoskeleton: a randomized controlled multicenter trial. Front. Neurosci. **11**, 400 (2017). https://doi.org/10.3389/fnins.2017.00400

M. Jeannerod, The representing brain: neural correlates of motor intention and imagery. Behav. Brain Sci. **17**(2), 187–202 (1994)

M. Jeannerod, Neural simulation of action: a unifying mechanism for motor cognition. Neuroimage **14**, S103–S109 (2001)

P. Langhorne, F. Coupar, A. Pollock, Motor recovery after stroke: a systematic review. Lancet Neurol. **8**(8), 741–754 (2009)

OA Mokienko, AV Chervyakov, SN Kulikova, PD Bobrov, LA Chernikova, AA Frolov, MA Piradov, Increased motor cortex excitability during motor imagery in brain-computer interface trained subjects. Front. Comput. Neurosci. **7**, 168 (2013). https://doi.org/10.3389/fncom.2013.00168

T. Ono, K. Shindo, K. Kawashima, N. Ota, M. Ito, T. Ota, M. Mukaino, T. Fujiwara, A. Kimura, M. Liu, Brain-computer interface with somatosensory feedback improves functional recovery from severe hemiplegia due to chronic stroke. Front. Neuroeng. **7**, 19 (2014)

S. Paolucci, G. Antonucci, M.G. Grasso, D. Morelli, E. Troisi, P. Coiro, M. Bragoni, Early versus delayed inpatient stroke rehabilitation: a matched comparison conducted in Italy. Arch. Phys. Med. Rehabil. **81**(6), 695–700 (2000)

A. Pollock, S.E. Farmer, M.C. Brady, P. Langhorne, G.E. Mead, J. Mehrholz, F. van Wijck, Interventions for improving upper limb function after stroke. Stroke **46**(3), e57–e58 (2015)

A. Ramos-Murguialday, D. Broetz, M. Rea, L. Laer, O. Yilmaz, F.L. Brasil, G. Liberati, M.R. Curado, E. Garcia-Cossio, A. Vyziotis, W. Cho, M. Agostini, E. Soares, S. Soekadar, A. Caria, L.G. Cohen, N. Birbaumer, Brain-machine interface in chronic stroke rehabilitation: a controlled study. Ann. Neurol. **74**(1), 100–108 (2013). https://doi.org/10.1002/ana.23879

J. Sanford, J. Moreland, L.R. Swanson, P.W. Stratford, C. Gowland, Reliability of the Fugl-Meyer assessment for testing motor performance in patients following stroke. Phys. Ther. **73**(7), 447–454 (1993)

A. Solodkin, P. Hlustik, E.E. Chen, S.L. Small, Fine modulation in network activation during motor execution and motor imagery. Cereb. Cortex **14**(11), 1246–1255 (2004)

A.G. Thrift, D.A. Cadilhac, T. Thayabaranathan, G. Howard, V.J. Howard, P.M. Rothwell, G.A. Donnan, Global stroke statistics. Int. J. Stroke **9**(1), 6–18 (2014)

# Hearables: In-Ear Multimodal Brain Computer Interfacing

**Metin C. Yarici, Harry J. Davies, Takashi Nakamura, Ian Williams, and Danilo P. Mandic**

**Abstract** The Brain Computer Interface (BCI) of the near future must be suitable for widespread use in real-world environments. As such, it will be robust, portable, user friendly and discreet—and ideally wearable. In addition, for 'affective' functionality, standard electroencephalogram (EEG) based BCI needs to be augmented with sensors for other physiological modalities. Our generic 'Hearables' earpiece, equipped with miniature multimodal sensors, provides such a multimodal solution for reliable measurement of both neural activity and vital signs. Real-world viability is demonstrated through single-channel, multimodal digital noise removal in the EEG, standard BCI responses and more than 100 h of out-of-clinic sleep analysis. The benefits of collocated, multimodal sensing of the neural function and vital signs within our Hearables are demonstrated to extend beyond the enhancement of current BCIs, and into BCI-enabled eHealth. Finally, the advantages of our device are validated in an 'affective' BCI setting—where both the mental and physical state of the user is integrated through simultaneous monitoring of brain and vital functions.

**Keywords** Affective brain-computer interface · Multimodal vital signs monitoring · Hearables · De-noising · Ear-EEG

## 1 Introduction

The application of BCI has significantly advanced since its inception as a communication or movement pathway for the severely disabled (Vidal 1973). Modern applications include BCI for entertainment, education, healthcare and workplace environments (Blankertz et al. 2010; Valeriani et al. 2017; Schuller 2017), through real-time, closed-loop, affective brain monitoring of healthy and non-healthy individuals alike. Almost all modern applications naturally require functionality outside the specialist

M. C. Yarici (✉) · H. J. Davies · T. Nakamura · I. Williams · D. P. Mandic
Department of Communications and Signal Processing, Electronic and Electrical Engineering,
Imperial College London, South Kensington Campus, London, UK
e-mail: metin.yarici16@imperial.ac.uk

© The Author(s), under exclusive license to Springer Nature Switzerland AG 2021
C. Guger et al. (eds.), *Brain-Computer Interface Research*,
SpringerBriefs in Electrical and Computer Engineering,
https://doi.org/10.1007/978-3-030-60460-8_7

79

setting, in the natural setting of the application. Of the numerous current developments in the field of real-world BCI, EEG-based devices are closest to leaving the laboratory setting (Casson 2019). However, conventional systems remain bulky, cumbersome and primarily operate in the specialist setting. In addition, for affective functionality, wherein a profile of the user's mental and physical state (e.g. their level of stress, fatigue or motivation) is integrated into the interface (Liu et al. 2010; Muhl et al. 2011), standard BCI devices need to be augmented with additional sensors for monitoring multiple vital signs, such as the electrocardiogram (ECG), respiration and blood oxygenation (Pantelopoulos and Bourbakis 2010; Xu et al. 2011; Troster 2005). Therefore, any future widely applicable EEG-based BCI should satisfy the following design requirements:

R1:   Portable, lightweight and easily carried,
R2:   Discreet, worn without attracting attention,
R3:   Robust, operated in real-world environments,
R4:   Prolonged, functional over hours and days,
R5:   Unobtrusive, compatible with the user's activity,
R6:   Multimodal, simultaneous monitoring of brain and vital signs activity.

## 2   Generic Multimodal Earpiece

The latest portable EEG headsets can offer up to 32 channels of continuous recording over a 24 h period (Neurosky.com/; www.emotiv.com/epoc). However, their use outside the specialist setting is limited by their stigmatising and obtrusive nature (Waterhouse 2003). In 2010, we pioneered ear-EEG as an ultra-wearable alternative to scalp-based portable EEG systems by utilising discreet and practical ear canal and intra-auricular electrodes embedded inside a hard shell earpiece to record the first BCI-related EEG signals (Looney et al. 2012; Looney and Mandic 2014). In the ensuing years, we have witnessed the introduction of numerous ear-EEG systems, in the context of BCI (Bleichner and Debener 2017), as well as objective audiometry (Christensen et al. 2018a, b), sleep analysis (Mikkelsen et al. 2019a, b; Nakamura et al. 2020; Alqurashi et al. 2019; Looney et al. 2016), cognitive load assessment (Wascher et al. 2019) and biometric authentication (Nakamura et al. 2018). However, early designs require a costly and time-consuming process to manufacture custom-fit earpieces, and, as a result of their rigid structure, are highly susceptible to motion artefacts. In addition, our initial designs, like the range of ear-EEG devices proposed thus far (Liu et al. 2010; Muhl et al. 2011; Pantelopoulos and Bourbakis 2010; Xu et al. 2011; Troster 2005; Waterhouse 2003; Looney et al. 2012; Neurosky.com/; www. emotiv.com/epoc), do not provide innovative methods to improve the reliability of naturally compromised wearable EEG sensing, or take into consideration the importance of achieving simultaneous brain and vital signs monitoring. Our subsequent device (Goverdovsky et al. 2017) comprises three key elements: generic viscoelastic substrate, flexible electrophysiology electrode and MMS, each of which provides solutions to the requirements R1:R6 above.

**Fig. 1** Construction of the multimodal in-ear sensing device. **a** Detailed structure of the device, showing the placement of the microphone and the electrode on the substrate. **b** Construction of the multimodal sensor underneath one of the textile electrodes. **c** Completed earpiece with electrodes and inward-facing microphone visible. **d** Placement of the earpiece in the user's ear

The substrate is a viscoelastic foam which, by virtue of its mechanical properties (significant relaxation and minimal outward pressure upon rapid compression Goverdovsky et al. 2015), provides passive mitigation of artefacts by absorbing energy that would otherwise create movement between the electrode and the surface of the skin, the so-called motion artefact. In our previous work, we demonstrated the foam's ability to absorb sufficient energy to completely eliminate the strong blood vessel-pulsation motion artefact in the ear-EEG signal (Goverdovsky et al. 2016). Simultaneously, the substrate provides a generic (universal), comfortable and secure fit that enables the user to move freely without the device dislodging. The second key element is a flexible capacitive electrophysiology electrode (Goverdovsky et al. 2016) which compliments the substrate's generic nature by conforming to any shape ear canal. The flexible electrode is constructed by securing a $4 \times 7$ mm strip of conductive textile to the surface of the substrate (see Fig. 1a–c). Electrical connections are provided through conductive yarn that is woven into the textile and passed through the body of the substrate. The electrode requires only small amounts of saline solution to provide prolonged, robust and low contact impedance with the ear canal surface (see Fig. 3a). The third key component is our MMS, which provides both EEG de-noising and vital signs monitoring capabilities. The sensor is constructed by mechanically coupling a microelectromechanical (MEMS) microphone to the bottom surface of one of the textile electrodes (Goverdovsky et al. 2015) (see Fig. 1a, b). In this way, once the device is placed inside the ear canal, mechanical interference stemming from motion at the recording site will be faithfully recorded by the MEMS microphone. As such, the microphone's signal can be used in digital noise removal of the motion artefact on the electrical signal recorded by the flexible electrode. Note that our MMS does not require extra room for artefact detection, in contrast with other proposed sensors for electrophysiological recording systems (Ko et al. 2012; Gibbs et al. 2005), and, unlike popular algorithmic methods for noise removal, does not require multi-channel data (Blankertz et al. 2010); this is a must in the wearable scenario where the use of fewer channels in a confined area is preferable. In addition, serving a dual functionality, the MMS microphone detects other acoustic signals that propagate through the body to the ear, including respiration, speech, and blood vessel pulsation (note that this is simultaneously a mitigated artefact in the EEG electrode's signal and a faithful proxy of heart rate in the microphone's signal).

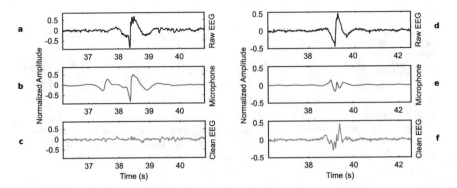

**Fig. 2** De-noising of ear-EEG from mechanical jaw clench artefacts. **a–c** Best-case de-noising scenario for which the artefact measurement with the MMS microphone was accurate. **d–f** Worst-case accuracy of artefact measurement with the MMS microphone; note that the artefact was still reduced. **a, d** Raw EEG with a strong artefact. **b, e** Output of the microphone within the MMS. **c, f** De-noised EEG using MMS microphone signal as reference

## 3 MMS: EEG De-noising

As stated above, and comprehensively demonstrated in Goverdovsky et al. (2015), the multimodal earpiece, by the very nature of its substrate, provides passive mitigation of the noise component in the EEG stemming from the motion artefact. However, in less favourable scenarios, such noise is not completely diminished by the properties of the substrate alone, particularly during strong jaw clenches, as shown in Fig. 2a, d. The MMS provides a solution for such stronger artefacts, as typically occur in real-world environments. An example of MMS noise removal of strong jaw-clenching, where the mechanical signal accurately detects the artefact, is depicted in Fig. 2a–c. Note the significant reduction in amplitude of the jaw clench artefact in the ear-EEG recording. Figure 2d–f illustrates results from a less favourable scenario, in which the mechanical signal recorded by the microphone is compromised, but nevertheless the motion artefact is still visibly attenuated. Whilst current BCIs often discard considerable amounts of artefact polluted data, our multimodal device provides a robust and practical solution for mitigating this loss.

## 4 BCI Responses in Ear-EEG: Standard EP's and Alpha

The viability of long-term EEG monitoring in real-world environments is demonstrated in a study of EEG-electrode impedance stability. The device is then validated against Scalp-EEG for both auditory steady-state response (ASSR) and steady-state visually evoked potential (SSVEP), in addition to visual evoked potentials (VEP) and alpha band sensitivity, see Fig. 3. The ASSR at a 40 Hz modulation frequency was obtained for both ear-EEG, the mastoid (M1), and the central (Cz) scalp electrodes,

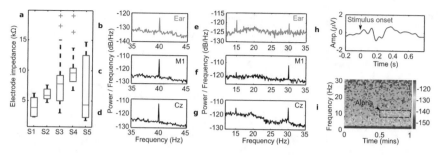

**Fig. 3** Standard BCI responses. **a** Electrode impedance for 5 subjects over 8 h. **b–d** ASSR and **e–g** SSVEP response measured from the ear, M1 and Cz scalp locations. **h** Visual evoked potential measured with the earpiece. **i** Alpha rhythm recorded from the ear electrodes; the subjects closed their eyes 30 s into the trial

and compared in Fig. 3b–d. Observe clear peaks at the modulation frequency for all the recording positions, with the signal to noise ratio (SNR) of the ear-EEG comparable to that from conventional on-scalp electrodes. The SSVEP was induced in EEG by presenting the subjects with an LED blinking at 15 Hz. As desired, a clear peak was observed at the stimulus frequency and its first harmonic at 30 Hz, as shown in Fig. 3e–g. We further demonstrate the functionality of our device to acquire transient neural responses by presenting the subjects with an LED switched ON for 200 ms and then fully OFF for 1800 ms. Figure 3h shows that the shape and timing of the VEP waveform, as measured from the generic earpiece, are a good match for the corresponding waveforms from scalp electrodes. Figure 3a demonstrates the robustness of the generic earpiece, as exemplified through impedance values of ear-EEG electrodes, measured at regular intervals over the course of an uninterrupted work-day (including activities such as walking, talking and eating) for five subjects. For all subjects, the electrode impedance maintained a low value (predominantly below 10 kΩ). Finally, the multimodal earpiece is shown to be capable of recording neural activity related to fatigue, drowsiness and sleepiness (the alpha band), observed clearly in Fig. 3i through a time-frequency representation of an alpha attenuation trial. Additional detail of the BCI response experiments can be found in (Goverdovsky et al. 2017).

## 5 Augmented BCI: Sleep Analysis via Ear-EEG

Sleep quality is an indicator of the state of body and mind, and an opportunity for BCI-enabled eHealth. Twenty-two healthy subjects took part in an overnight sleep study with simultaneous ear-EEG from the generic earpiece, and conventional full polysomnography (PSG) recordings (Nakamura et al. 2020). The ear-EEG and scalp-EEG data were used in automatic sleep stage prediction through supervised machine learning, whereby the PSG data, manually scored by a sleep clinician, served as ground truth. For rigorous real-world testing, each overnight recording was

**Table 1** Performance of the sensor in automatic sleep stage classification

| Sensor | No. of subjects included in analysis | Average data included in analysis per subject (hrs, %) | Algorithm performance | | | |
|--------|--------|--------|--------|--------|--------|--------|
| | | | Sensitivity | Precision | Accuracy | Kappa |
| Generic earpiece | 16/22 | 6.05, 76.8 | 72.28 | 72.64 | 74.1 | 0.61 |
| Scalp-EEG | 17/22 | 6.40, 81.7 | 56.76 | 80.66 | 85.9 | 0.79 |

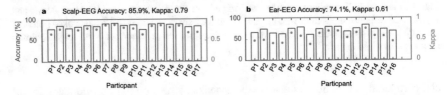

**Fig. 4** Classification accuracy (blue bars) and Cohen's kappa values (orange points) for individual participants, P1–17, in one overnight sleep trial. **a** Scalp-EEG based results, **b** ear-EEG based results

conducted in both an unsupervised (excluding equipment set-up and collection by a sleep clinician) and natural setting in the homes of the participants. The capability of the generic earpiece to record continuous and high-fidelity sleep-EEG is evidenced in Table 1. The number of participants and mean number of hours of clean ear-EEG and Scalp-EEG recorded data were, respectively, N = 16 (76.8%) and 6.05 h, and N = 17 (81.7%) and 6.40 h. The automatic sleep staging algorithm performed classification between wake, N1, N2, N3 and REM periods and obtained classification accuracy of 74.1% with a corresponding Cohen's $\kappa$ value of 0.61 (Substantial Agreement) when using data from the generic earpiece (Fig. 4).

## 6 Continuous Brain and Vital Signs Monitoring

BCI applications have typically focused on volitional control and superliminal feedback. However, just as in interpersonal relationships, subconscious communication and context are vital. Understanding the mental and physical state of the user can radically alter the relationship and effectiveness of the machine interface. In addition to EEG, the multimodal sensor detects blood vessel pulsation in the ear canal—a mechanical plethysmogram (MPG)—and in our latest implementations this has been combined with a reflective PPG sensor. The raw waveforms, and the subsequent HRV stress analysis for the ear-PPG, benchmarked against gold-standard chest-ECG HRV, are displayed, respectively, in Fig. 5a and d. Observe close matching of the ear-PPG to benchmark chest-ECG in HRV frequency analysis. In another modality supported by

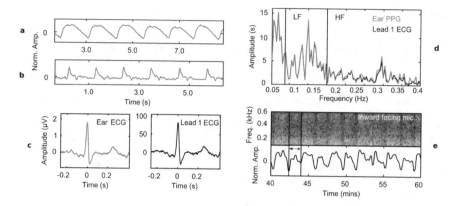

**Fig. 5** Augmented sensing for BCI: Ear **a** photo- and **b** mechanical-plethysmogram raw waveforms. **c** Ear- and Lead- 1 Chest-ECG waveforms from one subject. **e** Spectrogram of a breathing signal recorded from the inward facing microphone and normalised amplitude of the spectrogram, with one of the single respiratory periods indicated by the black lines and arrows

the earpiece, the MMS combines mechanical and electrical signals to obtain an ear-ECG, which has been comprehensively validated as an equivalent to standard Lead 1 chest-ECG (von Rosenberg et al. 2017). In addition, for blood pressure monitoring, the pulse-arrival time (PAT) is also reliably extracted through combining information from the mechanical/optical and electrical signals detected by our device (Goverdovsky et al. 2015, 2017; von Rosenberg et al. 2017). Another important feature of affective monitoring is respiration analysis (Picard 2000). The inward-facing microphone in our device detects endogenous acoustic (bone-conduction) signals that preserve respiration patterns (as well as speech Goverdovsky et al. 2017), as can be seen in a time-frequency representation of the acoustic signal depicted in Fig. 5e.

# 7 Summary

Stress, emotions and fatigue have major roles in determining the effectiveness of a BCI, while real-world functionality is key for future uses of the technique. To address this issue, we have demonstrated the potential of multimodal, integrated in-ear sensing by validating the so-called Hearables in real-world and affective BCI applications. Salient features of our multimodal BCI include single-channel enabled digital noise removal, standard BCI response detection and continuous out-of-lab sleep staging. By taking account of closed loop synchronisation between body and mind, we have demonstrated conclusively the feasibility of augmented and user-affected BCIs in the community via Hearable sensing.

# References

Y. Alqurashi et al., The efficacy of a novel in-ear electroencephalography (EEG) sensor to measure overnight sleep in healthy participants. Am. J. Respir. Crit. Care Med., **199** (2019)

B. Blankertz et al., The Berlin brain-computer interface: non-medical uses of BCI technology. Front. Neurosci. **4**, 198 (2010)

M.G. Bleichner, S. Debener, Concealed, unobtrusive ear-centered eeg acquisition: cEEGrids for transparent EEG. Front. Hum. Neurosci. **11**, 163 (2017)

A.J. Casson, Wearable EEG and beyond. Biomed. Eng. Lett. **9**(1), 53–71 (2019)

C.B. Christensen et al., Ear-EEG-based objective hearing threshold estimation evaluated on normal hearing subjects. IEEE Trans. Biomed. Eng. **65**(5), 1026–1034 (2018a)

C.B. Christensen et al., Toward EEG-assisted hearing aids: objective threshold estimation based on ear-EEG in subjects with sensorineural hearing loss. Trends Hear. **22**, 2331216518816203 (2018b)

*EEG—ECG—Biosensors.* Available: Neurosky.com/

P.T. Gibbs, L.B. Wood, H.H. Asada, Active motion artifact cancellation for wearable health monitoring sensors using collocated MEMS accelerometers. Smart Struct. Mater. 2005: Sens.S Smart Struct. Technol. Civ., Mech., Aerosp., Pts 1 and 2 **5765**, 811–819 (2005)

V. Goverdovsky et al., Co-located multimodal sensing: a next generation solution for wearable health. IEEE Sens. J. **15**(1), 138–145 (2015)

V. Goverdovsky et al., In-ear EEG from viscoelastic generic earpieces: robust and unobtrusive 24/7 monitoring. IEEE Sens. J. **16**(1), 271–277 (2016)

V. Goverdovsky et al., Hearables: multimodal physiological in-ear sensing. Sci. Rep. **7**, 6948 (2017)

B. Ko et al., Motion artifact reduction in electrocardiogram using adaptive filtering based on half cell potential monitoring. *2012 Annual International Conference of the IEEE Engineering in Medicine and Biology Society (EMBC)* (2012), pp. 1590–1593

Y. Liu, O. Sourina, M.K. Nguyen, Real-time EEG-based human emotion recognition and visualization. (2010)

D. Looney, D. Mandic, Ear-EEG: user-centered and wearable BCI, in *Brain-Computer Interface Research* (Springer, 2014)

D. Looney et al., The in-the-ear recording concept user-centered and wearable brain monitoring. IEEE Pulse **3**(6), 32–42 (2012)

D. Looney et al., Wearable in-ear encephalography sensor for monitoring sleep preliminary observations from nap studies. Ann. Am. Thorac. Soc. **13**(12), 2229–2233 (2016)

K.B. Mikkelsen et al., Accurate whole-night sleep monitoring with dry-contact ear-EEG. Scientific Reports **9**, 16824 (2019a)

K.B. Mikkelsen et al., Machine-learning-derived sleep-wake staging from around-the-ear electroencephalogram outperforms manual scoring and actigraphy. J. Sleep Res. **28**(2), UNSP e12786 (2019b)

C. Muhl et al., Affective brain-computer interfaces (aBCI 2011). Affect. Comput. Intell. Interact., Pt Ii **6975**, 435 (2011)

T. Nakamura, V. Goverdovsky, D.P. Mandic, In-ear EEG biometrics for feasible and readily collectable real-world person authentication. IEEE Trans. Inf. Forensics Secur. **13**(3), 648–661 (2018)

T. Nakamura et al., Hearables: automatic overnight sleep monitoring with standardized in-ear EEG sensor. IEEE Trans. Biomed. Eng. **67**(1), 203–212 (2020)

A. Pantelopoulos, N.G. Bourbakis, A survey on wearable sensor-based systems for health monitoring and prognosis. IEEE Trans. Syst. Man Cybern. Part C—Appl. Rev. **40**(1), 1–12 (2010)

R.W. Picard, *Affective Computing* (MIT Press, 2000)

B. Schuller, Can affective computing save lives? Meet mobile health. Computer **50**(5), 13 (2017)

G. Troster, The agenda of wearable healthcare. Yearbook Med. Inform. **1**, 125–138 (2005)

D. Valeriani, C. Cinel, R. Poli, Group augmentation in realistic visual-search decisions via a hybrid brain-computer interface. Sci. Rep. **7**, 7772 (2017)

J.J. Vidal, Toward direct brain-computer communication. Annu. Rev. Biophys. Bioeng. **2**, 157–180 (1973)

W. von Rosenberg et al., Hearables: feasibility of recording cardiac rhythms from head and in-ear locations. R. Soc. Open Sci. **4**(11), 171214 (2017)

E. Wascher et al., Evaluating mental load during realistic driving simulations by means of round the ear electrodes. Front. Neurosci. **13**, 940 (2019)

E. Waterhouse, New horizons in ambulatory electroencephalography. IEEE Eng. Med. Biol. Mag. **22**(3), 74–80 (2003)

*Wireless EEG Headset.* Available: www.emotiv.com/epoc

P. Xu, X. Tao, S. Wang, Measurement of wearable electrode and skin mechanical interaction using displacement and pressure sensors. (2011)

# Power Modulations of Gamma Band in Sensorimotor Cortex Correlate with Time-Derivative of Grasp Force in Human Subjects

Tianxiao Jiang, Priscella Asman, Giuseppe Pellizzer, Dhiego Bastos, Shreyas Bhavsar, Sudhakar Tummala, Sujit Prabhu, and Nuri F. Ince

**Abstract** Grasping objects of different size and weight is one of the most important hand functions in our daily lives. For this reason, a hand neuroprosthetic needs to be able to perform it with high accuracy. Previous brain-machine interface (BMI) studies often focused on decoding the kinematic part of the grasp such as individual finger position or velocity. Less is known about the kinetic part such as the generation and maintenance of grasp force. In this study, we recorded intraoperative high-density electrocorticography (ECoG) from the sensorimotor cortex of four patients while they executed a voluntary isometric hand grasp during awake surgeries. They were instructed to squeeze a hand-held dynamometer and maintain the grasp for 2–3 s before relaxing. We studied the power modulations of the neural oscillations during the whole time-course of the grasp including onset, hold, and offset phases. Phasic event-related desynchronization (ERD) in the low-frequency band (LFB) from 8 to 32 Hz and event-related synchronization (ERS) in the high-frequency band (HFB) from 60 to 200 Hz were observed at grasp onset and offset. However, during the holding period, the magnitude of LFB-ERD and HFB-ERS decreased near or at the baseline level. More importantly, we found that the fluctuations of HFB-ERS primarily, and of LFB-ERD to a lesser extent, correlated with the time-course of the first time-derivative of force (yank), rather than with force itself. To the best of our knowledge, this is the first study that establishes such a correlation. These results have fundamental implications for the decoding of grasp in brain oscillatory activity-based neuroprosthetics.

T. Jiang · P. Asman · N. F. Ince (✉)
Department of Biomedical Engineering, University of Houston, Houston, TX, USA
e-mail: nfince@uh.edu

G. Pellizzer
Departments of Neurology and Neuroscience, University of Minnesota, Minnesota, MN, USA

D. Bastos · S. Tummala · S. Prabhu
Department of Neurosurgery, University of Texas MD Anderson Cancer Center, Houston, TX, USA

S. Bhavsar
Department of Anesthesiology, University of Texas MD Anderson Cancer Center, Houston, TX, USA

© The Author(s), under exclusive license to Springer Nature Switzerland AG 2021
C. Guger et al. (eds.), *Brain-Computer Interface Research*,
SpringerBriefs in Electrical and Computer Engineering,
https://doi.org/10.1007/978-3-030-60460-8_8

**Keywords** Electrocorticography (ECoG) · Brain-machine interface (BMI) · Event-related desynchronization (ERD) · Even-related synchronization (ERS) · Neuroprosthetics · Grasp force

# 1 Motivation

Brain-machine interfaces (BMI) provide a way to restore motor function by decoding signals directly from the sensorimotor area of the brain (Collinger et al. 2013; Hochberg et al. 2012; Wang et al. 2013; Wodlinger et al. 2015; Yanagisawa et al. 2011, 2012). While grasping and holding objects are frequently executed in daily life, most BMI studies were focused on movement kinematics, such as individual finger position and velocity (Acharya et al. 2010; Branco et al. 2017; Chen et al. 2014; Flint et al. 2017; Hochberg et al. 2012; Hotson et al. 2016; Kubanek and Schalk 2014; Miller et al. 2009). However, the significance of dexterous control of exerted forces upon different objects, such as a cell phone versus an egg, can be easily overlooked. In order to establish a hand neuroprosthetic that can replicate natural hand function, it is crucial to understand the neural correlates of force control during a sustained grasp.

Previous studies have found a strong correlation between the activity of the flexor digitorum profundus finger muscle and the power in the delta (1.4–4 Hz) and gamma (50–90 Hz) subbands (Chen et al. 2014; Flint et al. 2014; Shin et al. 2012). However, these studies either investigated grasp force and relaxation without the grasp force being held for a prolonged period of time or analyzed only the onset (squeeze) phase of the grasp. In addition, the decoding algorithms used in these studies assumed a linear relation between neural oscillatory activity and grasp force. We hypothesized that the restriction of grasp force to brief grasp/hand posture changes masks the true nature of the relation between neural oscillations and grasp force.

To validate this hypothesis, we recorded high-density electrocorticography (ECoG) over the sensorimotor areas of four patients during awake craniotomies where each patient was instructed to perform an isometric grasp task. We observed phasic low-frequency band (8–32 Hz) event-related desynchronization (LFB-ERD) and high-frequency band (60–200 Hz) event-related synchronization (HFB-ERS) patterns at the grasp onset and offset, but not during the hold period (Chen et al. 2014; Ince et al. 2010; Jiang et al. 2018; Jiang et al. 2017a; Miller et al. 2007, 2010; Pfurtscheller and Lopes da Silva 1999; Ray et al. 2008; Sanes and Donoghue 1993; Su and Ojemann 2013; Tzagarakis et al. 2010). With our results, we show that the separation between the onset (squeeze), hold (steady force), and offset (relaxation) phases of the grasp is crucial for elucidating the relation between the time-course of grasp force and oscillatory neural activity. We found that the motor oscillatory activity during sustained grip tasks correlated strongly with the first time-derivative of force (or yank) rather than with force itself. Additional details of the recording methods, data processing steps, and results are available in (Jiang et al. 2020).

## 2  Methods

### 2.1  Subject Recruitments

We recruited four patients (2 females and 2 males; ages within 40–65 years) at the University of Texas MD Anderson Cancer Center (Houston, TX), who were diagnosed with a brain tumor and scheduled for a craniotomy in the vicinity of the sensorimotor area. Behavioral task examinations were conducted a day prior to the surgery to exclude motor deficits and to acquaint all patients to the experimental paradigm. The study protocol was reviewed and approved by the Institutional Review Boards (IRB) of the MD Anderson Cancer Center and the University of Houston. Informed consent was obtained from all four patients before their participation in the study in accordance with the Declaration of Helsinki.

### 2.2  ECoG Recordings

The surgery was performed using the sleep-awake-sleep anesthetic technique (Huncke et al. 1998). After the dura was opened in all 4 patients (P1–4), motor and sensory cortices were identified with cortical stimulation and/or median somatosensory evoked potentials phase reversal technique (MSSEP-PRT) (de Witt Hamer et al. 2012; Giussani et al. 2010; Goldring 1978; Goldring and Gregorie 1984; Korvenoja et al. 2006; Sheth et al. 2013; Simon et al. 2012, 2014). Functional mapping with intraoperative ECoG was performed during the awake craniotomies. A customized 128 channel grid (16 × 8, 1.17 mm contact exposure and 4 mm spacing, platinum, Ad-Tech, Michigan, MI) was used for P1, whiles customized 192 channel grids (16 × 12, 1 mm contact exposure and 3 mm spacing, platinum, PMT, Chanhassen, MN) were used for P2, P3, and P4, Fig. 1a, b. The portable data acquisition system setup used in this study, shown in Fig. 1d, was specifically designed for intraoperative neural and behavioral recordings (Jiang et al. 2017b).

During the awake surgery, the patients were asked to execute a sustained grasp task and maintain the grasping force for at least two seconds before relaxing it. An inter-trial interval of approximately three seconds was maintained between the instruction of hand relaxation and the consecutive hand squeezing instruction. The grasp force level was measured with an analog hand dynamometer (Vernier HD-BTA) and digitized using a microcontroller (Teensy 3.1) at 100 Hz and 12-bit A/D resolution before being transferred to a dedicated laptop over the User Datagram Protocol using in-house custom-made software. Hand movements were monitored using a high-definition webcam (Logitech HD C270). The neural data and forearm bipolar electromyogram (EMG) were recorded with a 256-channel clinical bio-amplifier (gHIamp, g.tec medical engineering GmbH, Graz, Austria) at 2.4 kHz. All behavioral

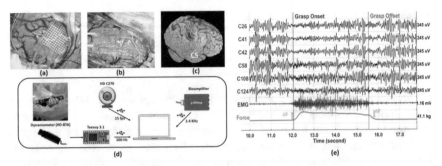

**Fig. 1** **a** High-density ECoG used in P1 (16 × 8, 4 mm spacing). **b** High-density ECoG used in P2, P3 and P4 (16 × 12, 3 mm spacing). The central sulcus is marked with a white line in each case. **c** 3D cortical mesh obtained from patient MRI with electrode placement showing active channels. **d** Recording system setup used in the operating room. The subject squeezed the force sensor for at least 2–3 s. The sensor output was acquired with a microcontroller and transmitted to a nearby laptop to be synchronized with the neural data streaming from the bioamplifier. **e** Raw ECoG from selected channels is shown along with force sensor readings and EMG. We can notice that the oscillatory activity in beta range (13–30 Hz) was suppressed at the grasp onset and that, although the subject maintained the grip force, the amplitude of the oscillations recovered during the hold period. During grasp offset, the oscillatory activity was suppressed again. High frequency gamma activity can be observed in channels C58, C108, and C124. EMG and force traces are shown at the bottom (Figure modified from Jiang et al. 2020)

and neural data were acquired, synchronized, and visualized in real-time intraoperatively using Simulink/Matlab and gHISYS block sets (g.tec medical engineering GmbH, Graz, Austria).

## 2.3 Preprocessing

All ECoG recordings were visually examined to remove corrupted channels and artifacts. The power line noise at 60 Hz and its harmonics were removed via a series of second-order infinite impulse response (IIR) notch filters. We used the minimum acceleration criterion with constraints (MACC) method to detect the beginning of grasp onset and offset on the root mean square (RMS) of EMG (Botzer 2009; Jiang et al. 2018) for event aligned analysis. As in our recent study (Jiang et al. 2017a), we executed a visual inspection using synchronized video, EMG, and force sensors to ensure that a detected onset point was a real grasp execution onset rather than an artifact. Trials with clean ramping EMG at grasp onset and without spontaneous EMG bursts over the 1.5 s baseline period before grasp onset were used. Figure 1e shows an epoch of ECoG data, forearm EMG, synchronized force, as well as grasp onset-offset from P2 data.

## 2.4 Time-Frequency Analysis

Data segments around grasp onset and offset ($-1.5$–$1.5$ s) were extracted and time-frequency analysis was performed at each channel using short-time Fourier transform (STFT). Specifically, the power spectral density (PSD) was estimated using a 512-sample Hanning window. The window was shifted with a 480-sample overlap for a smooth temporal transition. Smoothed spectrograms of grasp onset and offset for each channel were obtained by averaging the spectrograms across trials. The average spectrograms for grasp onset and offset were normalized using the same baseline PSD and transformed into a dB scale to yield the centered spectrograms.

## 2.5 Event-Related Synchronization and Desynchronization Around Grasp Onset, Hold and Offset

The original signal was bandpass filtered in the low-frequency band (LFB: 8–32 Hz) and the high-frequency band (HFB: 60–200 Hz) using a second-order Butterworth IIR zero-phase filter (forward and backward). The filtered signals were squared to compute the power traces for LFB and HFB. The temporal evolution of ERD and ERS was computed by normalizing the power traces against their respective baseline power preceding the grasp onset.

For grasp onset, ERD and ERS were computed from $-0.1$ to $0.7$ s around the grasp onset. For grasp offset, they were estimated from the 0.8 s of data following immediately after grasp offset. For the hold phase, we used 1–1.5 s of data segments in which the grip force reached a plateau. These segments were at least 1 s away from the grasp onset and offset.

The statistical significance of ERD/ERS at each channel was tested using a one-tailed Student's $t$-test with a significance threshold $p$-value of 0.05 and corrected for multiple comparisons using the False Discovery Rate (FDR) method at the level of 0.05 (Genovese et al. 2002). In order to select the most robust ERD and ERS events, we selected channels that had significant changes of at least 25% from baseline in either grasp onset or offset phase as previously done in (Jiang et al. 2018).

## 2.6 Electrode Localization

The grid localization was determined by the neurosurgeons based on the coregistration of the intraoperative photograph of the cortex and the preoperative MRI scan of the brain using bio-landmarks such as blood vessels, sulci, and gyri, Fig. 1c.

## 2.7   Data Analysis

For each patient, we performed a cross-correlation analysis between the average time-series of LFB-ERD/HFB-ERS and the first time-derivative of grasp force (yank). The time-series were averaged across trials to improve the signal to noise ratio and yield a smooth estimation of the lag related to maximum/minimum cross-correlation coefficients. After that, the Pearson correlation coefficient was computed for individual trials by shifting LFB-ERD/HFB-ERS according to the identified lag from the averaged data. The correlation around grasp onset and offset were analyzed separately using the ECoG channels associated with significant LFB-ERD or HFB-ERS activations and using channels anterior or posterior to the central sulcus. Specifically, $-1$ to 1.5 s of data segment around grasp onset/offset was extracted to include the hold phase.

## 3   Results

### 3.1   Time-Frequency Analysis of Individual Patients

Average time-frequency maps for channels anterior and posterior to the central sulcus around grasp onset and offset are shown for each patient in Fig. 2. For all four patients, two distinct power modulations, one in LFB (8–32 Hz, blue) and the other in HFB (60–200 Hz, red), can be observed at both grasp onset and offset. Note that, although a sustained force level was maintained throughout the trial, the ERD and ERS magnitude decreased and returned towards baseline between grasp onset and offset. Overall, LFB-ERD lasted longer than HFB-ERS for all patients. LFB-ERD in P1 and P2 decreased close to baseline level after grasp onset, whereas for P3 and P4, LFB-ERD lasted throughout the hold period with only a slightly decreased magnitude. Figure 2 also shows channels that had significant activations at either grasp onset, offset, or both on the ECoG grids for each patient.

### 3.2   Temporal Evolution of ERD/ERS Regarding Force and Force Yank

The temporal evolution of average LFB-ERD (blue) and HFB-ERS (red) around grasp onset and offset is shown together with the average force trace (black) and force yank (dF) (red) in Fig. 3. In all patients, the power of LFB decreased and HFB increased during the onset and offset phases of the grasp. Even though a sustained force level was maintained throughout the hold phase, the magnitude of LFB-ERD and HFB-ERS decreased and returned at or close to the baseline level. Both LFB-ERD and HFB-ERS preceded grasp onset by 100 ms to 300 ms. At grasp offset,

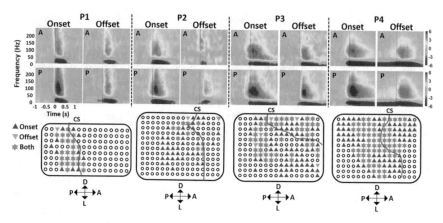

**Fig. 2** Centered time frequency maps averaged across significant channels anterior (A) or posterior (P) to the central sulcus are shown around grasp onset and offset for each patient. Each time-frequency map covered −1 to 1.25 s of the grasp onset/offset with a spectral range of 0 to 250 Hz. All maps are displayed from −6 to 6 dB. The channels used for averaging were marked on the grids below with upper triangles denoting significant channels during grasp onset, downward triangles denoting channels significant during grasp offset, and stars denoting channels significant during both grasp phases. The location of the central sulcus is marked by a red line on each electrode grid. The orientation of each grid is indicated at the bottom (A: anterior, P: posterior, D: dorsal, L: lateral) (Figure modified from Jiang et al. 2020. Supplementary Material)

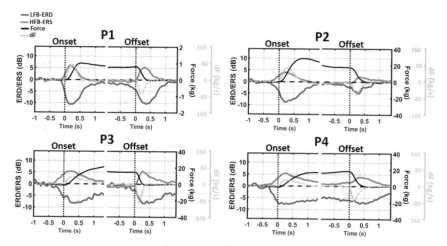

**Fig. 3** Temporal evolution of average LFB-ERD (blue), HFB-ERS (red), grip force (black), and the first time-derivative of force (dF, orange) around grasp onset and offset phases. The ERD/ERS were averaged across all significant channels and displayed from -14 to 14 dB for all patients while the scales of Force and dF are different across patients and provided on the right side of each figure. Both LFB-ERD and HFB-ERS started between 100 and 300 ms prior to the onset of the grasp (Figure modified from Jiang et al. 2020)

however, ERS generally started right at or after the beginning of the relaxation phase. In contrast, LFB-ERD generally started before the relaxation phase. The peaks of both LFB-ERD and HFB-ERS occurred after the peak of the force yank, that is, 200–300 ms after the start of the relaxation phase. Especially during grasp onset, the time-course of HFB-ERS closely matched the force yank for all patients. The similarities between ERD/ERS and force yank during grasp offset were not as high as during grasp onset. Overall, at grasp onset, the correlation between HFB-ERS and force yank ($r = 0.79 \pm 0.03$) was significantly stronger than with raw force ($r = 0.31 \pm 0.2$; paired $t(3) = 10.1$, $p = 0.002$). The correlation between HFB-ERS and force yank ($r = -0.61 \pm 0.1$) was slightly weaker at grasp offset and was not significantly different than the correlation with the raw force ($r = 0.63 \pm 0.1$; paired $t(3) = 0.23$, $p = 0.83$). At grasp onset and offset, the correlation between LFB-ERD and force yank was stronger (onset: $r = -0.6 \pm 0.08$; offset: $r = 0.35 \pm 0.17$) than the raw force (onset: $r = 0.44 \pm 0.05$; offset: $r = 0.24 \pm 0.4$), and weaker than to the HFB-ERS.

# 4    Discussion

## 4.1    Temporal Dynamics of LFB-ERD and HFB-ERS

Earlier ECoG BMI studies focusing on hand grasp have mainly investigated the decoding of movement kinematics, such as finger positions, grasp aperture, and velocity (Kubánek et al. 2009; Miller et al. 2009; Acharya et al. 2010; Pistohl et al. 2012; Nakanishi et al. 2014; Flint et al. 2017). The few studies that investigated the decoding of the kinetic aspects of hand grasp often used grasp tasks with relatively short (<0.1 s) or no explicit holding period (Chen et al. 2014; Flint et al. 2014). In this study, intraoperative high-density ECoG was recorded from the sensorimotor cortex of four patients while they were instructed to execute sustained hand grasps which lasted for at least 2–3 s. We observed phasic power modulations in ECoG subbands during the onset and offset phases. Although the subjects sustained the grasp, both gamma ERS and alpha/beta ERD diminished or vanished during the hold period.

To the best of our knowledge, the earliest evidence regarding phasic modulations in motor cortical oscillations was provided by Jasper and Penfield during the onset and offset phases of hand clenching (Jasper and Penfield 1949). A study of sustained hand grasp in non-human primates also showed distinct peaks of gamma modulations at both grasp onset and offset (Waldert et al. 2015). Both studies coincide with our observations in power modulation in LFB and HFB at movement onset and offset.

## 4.2 Connections to the Dynamics of the Afferent Systems

During grasp onset, finger movements and the pressure exerted on the hand dynamometer activate not only the muscle spindles but also the slowly-adapting, low-threshold mechanoreceptors, such as the Merkel cell-neurite complex found in the basal layer of the epidermis of fingers, as well as the slowly-adapting Ruffini endings broadly expressed in the dermis. The handling of the hand dynamometer also stimulates the rapidly-adapting Meissner corpuscles located in the dermal papillae of the glabrous skin as well as the rapidly-adapting Pacinian corpuscles, by way of skin deformation and indentation (Delhaye et al. 2018; Roudaut et al. 2012). It is likely that the late HFB-ERS at grasp offset is associated with the burst activity of fast adapting Meissner and Pacinian corpuscles due to the release of stimulation and skin deformation recovery. In recent work on the response to deep, light and soft touch (Kramer et al. 2019) with ECoG, elevations in high gamma (HG) power was shown within selected electrodes over the hand area of primary somatosensory cortex (S1), that lasted between around 300–500 ms, but extinguished prior to the end of the tactile stimulus. This coincides with our findings regarding the S1 high-gamma attenuation pattern during the hold period of a grasp task and Fig. 2 shows that high-intensity gamma response (50–150 Hz) peaked between 50 and 400 ms after grasp onset, within selected electrodes anterior and posterior to the central sulcus.

Interestingly, HFB-ERS was mostly at the lower end of the high gamma spectrum (60–200 Hz) during the hold period. Together with the phasic response of fast adapting receptors, it is likely that the fast firings generated by the slowly adapting Ruffini endings and Merkel cells contribute to the early high-intensity broadband gamma response that we observed (Fig. 2) by way of skin deformation, indentation, as well as pressure at the onset of the grasp (Delhaye et al. 2018; Kramer et al. 2019; Roudaut et al. 2012; Ryun et al. 2017). The broadband high-frequency ERS later reduces to a low-intensity gamma response at a lower frequency as time progresses, and this behavior can be related to the activation patterns of the slowly adapting receptors of the afferent system (Delhaye et al. 2018; Roudaut et al. 2012).

Both LFB-ERD and HFB-ERS started earlier than the grip onset, suggesting that both modulations are associated with the efferent movement command. Interestingly, while LFB-ERD started earlier, the HFB-ERS lagged the grip offset. Since LFB-ERD preceded both the onset and offset of the grasp and peaked after each, it is likely that the efferent system modulates the 8–32 Hz range at each phase, but this was later modulated by the afferent system. Consequently, the results suggest that the initiation and termination of grasp are associated with a distinct neural activity where the HFB-ERS represents the dynamics of the afferent and efferent systems at the grip onset, whereas HFB-ERS reflects mainly the afferent system during the offset that corresponds to the relaxation phase. The magnitude of LFB-ERD and HFB-ERS were weaker in the offset phase compared to the onset phase. The grasp onset phase does require more control than the relaxation phase, which may explain why there is a less clear signature of neural involvement before the relaxation phase.

## 4.3   Spatial Profile of HFB-ERS and LFB-ERD

We found that HFB-ERS was generally more extended, and of greater amplitude in posterior channels than anterior ones. In a recent study (Ryun et al. 2017), it was also discovered that HG activity in S1 was more dominant than in M1 during active, voluntary movement. Others have also confirmed that humans' sensory information is present in M1 recordings, in addition to motor responses in S1 (Sanes et al. 1995; Schieber and Hibbard 1993; Schroeder et al. 2017). In addition, M1 and S1 are reciprocally connected (Arce-McShane et al. 2016; Kunzle 1978). For these reasons, it may not be entirely surprising that voluntary movement-related oscillatory activity from M1 and S1 share similar characteristics and have relatively small quantitative differences.

## 4.4   Possible Challenges in ECoG Decoding for Sustained Hand Grasp

Published studies on decoding the force of hand grasp generally assumed a linear relation between brain oscillatory activity, such as beta and gamma-band power of ECoG or LFP, and grasp force (Chen et al. 2014; Flint et al. 2014; Milekovic et al. 2015; Tan et al. 2016). However, in this study, we show that the dynamics of LFB-ERD and HFB-ERS were more congruent with the first time-derivative of force rather than with force itself. In spite of a wide spectrum of force generated (1–30 kg), significant correlations between the time-course of LFB-ERD/HFB-ERS and the first time-derivative of force (yank) were found across all four patients. At grasp onset and offset, the correlation levels between HFB-ERS and force yank were significantly higher in posterior channels compared to anterior channels. Compared to HFB-ERS, LFB-ERD had weaker correlations with force yank. At grasp offset, the correlation between HFB-ERS/LFB-ERD and force yank was inverted and both neural activations lagged the force yank. Recently, Branco et al., also reported similar dynamics in high-frequency band suggesting nonlinear relationship between grasp force and neural modulations in sensorimotor cortex (Branco et al. 2019). In addition, the correlation between HFB-ERS and force yank was smaller for both anterior channels and posterior channels at grasp offset compared to grasp onset.

Moreover, the lack of HFB-ERS patterns forecasting the transition to the relaxation at the offset of the grasp will likely add challenges in decoding oscillatory neural activity involving sustained hand grasp. It is likely that information across multiple frequency bands and nonlinear dynamics of oscillatory activity need to be integrated to accurately decode the state transitions and the details of the hand grasp.

# 5   Conclusions

Understanding the neural encoding of force generation during a hand grasp task is important for the development of neuroprosthesis for common daily motor activities. In this study, we recorded high-density ECoG intraoperatively from the sensorimotor cortex of four patients while they executed a sustained hand grasp. Although the grasp force was maintained during hold, the magnitude of LFB-ERD and HFB-ERS decreased towards the baseline. Consistently in all patients, we show that the temporal dynamics of gamma ERS and beta ERD were correlated with the first time-derivative of force (yank) rather than with force itself. To the best of our knowledge, this is the first study that establishes such a correlation. These results have fundamental implications for the decoding of grasp in brain oscillatory activity-based neuroprosthetics. In general, due to the biphasic characters of HFB-ERS/LFB-ERD at grasp onset and offset, force decoding algorithms based on the cortical oscillatory activity should be carefully designed to preserve the memory of the system. Future strategies aimed at decoding sustained grasp force from subband modulations will need to model the first-time derivative of grasp force to develop useful neuroprostheses.

# References

S. Acharya, M.S. Fifer, H.L. Benz, N.E. Crone, N.V. Thakor, Electrocorticographic amplitude predicts finger positions during slow grasping motions of the hand. J. Neural. Eng. **7**(4), 046002 (2010). https://doi.org/10.1088/1741-2560/7/4/046002

F.I. Arce-McShane, C.F. Ross, K. Takahashi, B.J. Sessle, N.G. Hatsopoulos, Primary motor and sensory cortical areas communicate via spatiotemporally coordinated networks at multiple frequencies. Proc. Natl. Acad. Sci. U.S.A. **113**(18), 5083–5088 (2016). https://doi.org/10.1073/pnas.1600788113

L. Botzer, A simple and accurate onset detection method for a measured bell-shaped speed profile. Front. Neurosci. **3**(June), 1–8 (2009). https://doi.org/10.3389/neuro.20.002.2009

M.P. Branco, Z.V. Freudenburg, E.J. Aarnoutse, M.G. Bleichner, M.J. Vansteensel, N.F. Ramsey, Decoding hand gestures from primary somatosensory cortex using high-density ECoG. NeuroImage **147**, 130–142 (2017). https://doi.org/10.1016/j.neuroimage.2016.12.004

M.P. Branco, S.H. Geukes, E.J. Aarnoutse, M.J. Vansteensel, Z.V. Freudenburg, N.F. Ramsey. High-frequency band temporal dynamics in response to a grasp force task. J Neural Eng. **16**(5), 056009 (2019, Aug 6). https://doi.org/10.1088/1741-2552/ab3189

C. Chen, D. Shin, H. Watanabe, Y. Nakanishi, H. Kambara, N. Yoshimura, A. Nambu, T. Isa, Y. Nishimura, Y. Koike, Decoding grasp force profile from electrocorticography signals in non-human primate sensorimotor cortex. Neurosci. Res. **83**, 1–7 (2014). https://doi.org/10.1016/j.neures.2014.03.010

J.L. Collinger, B. Wodlinger, J.E. Downey, W. Wang, E.C. Tyler-Kabara, D.J. Weber, A.J.C. McMorland, M. Velliste, M.L. Boninger, A.B. Schwartz, High-performance neuroprosthetic control by an individual with tetraplegia. Lancet **381**(9866), 557–564 (2013). https://doi.org/10.1016/S0140-6736(12)61816-9

P.C. de Witt Hamer, S.G. Robles, A.H. Zwinderman, H. Duffau, M.S. Berger, Impact of intraoperative stimulation brain mapping on glioma surgery outcome: a meta-analysis. J. Clin. Oncol. **30**(20), 2559–2565 (2012). https://doi.org/10.1200/JCO.2011.38.4818

B.P. Delhaye, K.H. Long, S.J. Bensmaia, Neural basis of touch and proprioception in primate cortex. Compr. Physiol. **8**(4), 1575–1602 (2018). https://doi.org/10.1002/cphy.c170033

R.D. Flint, J.M. Rosenow, M.C. Tate, M.W. Slutzky, Continuous decoding of human grasp kinematics using epidural and subdural signals. J. Neural. Eng. **14**(1), 16005 (2017). http://stacks.iop.org/1741-2552/14/i=1/a=016005

R.D. Flint, P.T. Wang, Z.A. Wright, C.E. King, M.O. Krucoff, S.U. Schuele, J.M. Rosenow, F.P.K. Hsu, C.Y. Liu, J.J. Lin, M. Sazgar, D.E. Millett, S.J. Shaw, Z. Nenadic, A.H. Do, M.W. Slutzky, Extracting kinetic information from human motor cortical signals. NeuroImage **101**, 695–703 (2014). https://doi.org/10.1016/j.neuroimage.2014.07.049

C.R. Genovese, N.A. Lazar, T. Nichols, Thresholding of statistical maps in functional neuroimaging using the false discovery rate. NeuroImage **15**(4), 870–878 (2002). https://doi.org/10.1006/nimg.2001.1037

C. Giussani, F.E. Roux, J. Ojemann, E.P. Sganzerla, D. Pirillo, C. Papagno, Is preoperative functional magnetic resonance imaging reliable for language areas mapping in brain tumor surgery? Review of language functional magnetic resonance imaging and direct cortical stimulation correlation studies. *Neurosurgery* **66**(1), 113–120 (2010). https://doi.org/10.1227/01.NEU.0000360392.15450.C9

S. Goldring, A method for surgical management of focal epilepsy, especially as it relates to children. J. Neurosurg. **49**(3), 344–356 (1978). https://doi.org/10.3171/jns.1978.49.3.0344

S. Goldring, E.M. Gregorie, Surgical management of epilepsy using epidural recordings to localize the seizure focus. Review of 100 cases. J. Neurosurg. **60**(3), 457–466 (1984). https://doi.org/10.3171/jns.1984.60.3.0457

L.R. Hochberg, D. Bacher, B. Jarosiewicz, N.Y. Masse, J.D. Simeral, J. Vogel, S. Haddadin, J. Liu, S.S. Cash, P. van der Smagt, J.P. Donoghue, Reach and grasp by people with tetraplegia using a neurally controlled robotic arm. Nature **485**(7398), 372–375 (2012)

G. Hotson, D.P. McMullen, M.S. Fifer, M.S. Johannes, K.D. Katyal, M.P. Para, R. Armiger, W.S. Anderson, N.V. Thakor, B.A. Wester, N.E. Crone, Individual finger control of a modular prosthetic limb using high-density electrocorticography in a human subject. J. Neural. Eng. **13**(2), 26017 (2016). https://doi.org/10.1088/1741-2560/13/2/026017

K. Huncke, B. van de Wiele, E.H. Rubinstein, I. Fried, The asleep-awake-asleep anesthetic technique for intraoperative language mapping. Neurosurgery **42**(6), 1312–1316 (1998). https://doi.org/10.1097/00006123-199806000-00069

N.F. Ince, R. Gupta, S. Arica, A.H. Tewfik, J. Ashe, G. Pellizzer, High accuracy decoding of movement target direction in non-human primates based on common spatial patterns of local field potentials. PLoS ONE **5**(12), e14384 (2010). https://doi.org/10.1371/journal.pone.0014384

H. Jasper, W. Penfield, Electrocorticograms in man: effect of voluntary movement upon the electrical activity of the precentral gyrus. Archiv Für Psychiatrie Und Nervenkrankheiten **183**(1), 163–174 (1949). https://doi.org/10.1007/BF01062488

T. Jiang, T. Jiang, T. Wang, S. Mei, Q. Liu, Y. Li, X. Wang, S. Prabhu, Z. Sha, N.F. Ince, Characterization and decoding the spatial patterns of hand extension/flexion using high-density ECoG. IEEE Trans. Neural Syst. Rehabil. Eng. **4320**(c), 1–1 (2017a). https://doi.org/10.1109/TNSRE.2016.2647255

T. Jiang, H. Siddiqui, S. Ray, P. Asman, M. Ozturk, N.F. Ince, A portable platform to collect and review behavioral data simultaneously with neurophysiological signals, in *Proceedings of 39th Annual International Conference of the IEEE Engineering in Medicine and Biology Society, EMBS* (2017b), pp. 1784–1787. https://doi.org/10.1109/EMBC.2017.8037190

T. Jiang, S. Liu, G. Pellizzer, A. Aydoseli, S. Karamursel, P.A. Sabanci, A. Sencer, C. Gurses, N.F. Ince, Characterization of hand clenching in human sensorimotor cortex using high-, and ultra-high frequency band modulations of electrocorticogram. Front. Neurosci. **12** (February 2018). https://doi.org/10.3389/fnins.2018.00110

T. Jiang, G. Pellizzer, P. Asman, D. Bastos, S. Bhavsar, S. Tummala, S. Prabhu, N.F. Ince, Power modulations of ECoG alpha/beta and gamma bands correlate with time-derivative of force during hand grasp. Front. Neurosci. **14**, 100 (2020). https://doi.org/10.3389/fnins.2020.00100

A. Korvenoja, E. Kirveskari, H.J. Aronen, S. Avikainen, A. Brander, J. Huttunen, R.J. Ilmoniemi, J.E. Jääskeläinen, T. Kovala, J.P. Mäkelä, E. Salli, M. Seppä, Sensorimotor cortex localization: comparison of magnetoencephalography, functional MR imaging, and intraoperative cortical mapping. Radiology **241**(1), 213–222 (2006). https://doi.org/10.1148/radiol.2411050796

D.R. Kramer, M.F. Barbaro, M. Lee, T. Peng, G. Nune, C.Y. Liu, S. Kellis, B. Lee, Electrocortico-graphic changes in field potentials following natural somatosensory percepts in humans. Exp. Brain Res. **237**(5), 1155–1167 (2019). https://doi.org/10.1007/s00221-019-05495-1

J. Kubánek, K.J. Miller, J.G. Ojemann, J.R. Wolpaw, G. Schalk, Decoding flexion of individual fingers using electrocorticographic signals in humans. J. Neural Eng. **6**(6), 066001 (2009). https://doi.org/10.1088/1741-2560/6/6/066001

J. Kubanek, G. Schalk, NeuralAct: a tool to visualize electrocortical (ECoG) activity on a three-dimensional model of the cortex. Neuroinformatics **13**(2), 167–174 (2014). https://doi.org/10.1007/s12021-014-9252-3

H. Kunzle, Cortico-cortical efferents of primary motor and somatosensory regions of the cerebral cortex in Macaca fascicularis. Neuroscience **3**(1), 25–39 (1978). https://doi.org/10.1016/0306-4522(78)90151-3

T. Milekovic, W. Truccolo, S. Grun, A. Riehle, T. Brochier, Local field potentials in primate motor cortex encode grasp kinetic parameters. NeuroImage **114**, 338–355 (2015). https://doi.org/10.1016/j.neuroimage.2015.04.008

K.J. Miller, G. Schalk, E.E. Fetz, M. den Nijs, J.G. Ojemann, R.P.N. Rao, Cortical activity during motor execution, motor imagery, and imagery-based online feedback. Proc. Natl. Acad. Sci. **107**(9), 4430–4435 (2010). https://doi.org/10.1073/pnas.0913697107

K.J. Miller, S. Zanos, E.E. Fetz, M. den Nijs, J.G. Ojemann, Decoupling the cortical power spectrum reveals real-time representation of individual finger movements in humans. J. Neurosci. **29**(10), 3132–3137 (2009)

K.J. Miller, M. DenNijs, P. Shenoy, J.W. Miller, R.P.N. Rao, J.G. Ojemann, Real-time functional brain mapping using electrocorticography. Neuroimage **37**, 504–507 (2007)

Y. Nakanishi, T. Yanagisawa, D. Shin, C. Chen, H. Kambara, N. Yoshimura, N., et al., Decoding fingertip trajectory from electrocorticographic signals in humans. Neurosci. Res. **85**, 20–27 (2014). https://doi.org/10.1016/j.neures.2014.05.005

G. Pfurtscheller, F.H. Lopes da Silva, Event-related EEG/MEG synchronization and desynchro-nization: basic principles. Clin. Neurophysiol.: Off. J. Int. Fed. Clin. Neurophysiol. **110**(11), 1842–1857 (1999). http://www.ncbi.nlm.nih.gov/pubmed/10576479

T. Pistohl, A. Schulze-Bonhage, A. Aertsen, C. Mehring, T. Ball, Decoding natural grasp types from human ECoG. Neuroimage **59**, 248–260 (2012). https://doi.org/10.1016/j.neuroimage.2011.06.084

S. Ray, N.E. Crone, E. Niebur, P.J. Franaszczuk, S.S. Hsiao, Neural correlates of high-gamma oscillations (60–200 Hz) in macaque local field potentials and their potential implications in electrocorticography. J. Neurosci.: Off. J. Soc. Neurosci. **28**(45), 11526–11536 (2008). https://doi.org/10.1523/JNEUROSCI.2848-08.2008

Y. Roudaut, A. Lonigro, B. Coste, J. Hao, P. Delmas, M. Crest, Touch sense: functional organization and molecular determinants of mechanosensitive receptors. Channels (Austin, Tex.) **6**(4), 234–245 (2012). https://doi.org/10.4161/chan.22213

S. Ryun, J.S. Kim, H. Lee, C.K. Chung, Tactile frequency-specific high-gamma activities in human primary and secondary somatosensory cortices. Sci. Rep. **7**(1), 1–10 (2017). https://doi.org/10.1038/s41598-017-15767-x

J.N. Sanes, J.P. Donoghue, Oscillations in local field potentials of the primate motor cortex during voluntary movement. Proc. Natl. Acad. Sci. U.S.A. **90**(10), 4470–4474 (1993). https://www.ncbi.nlm.nih.gov/pubmed/8506287

J.N. Sanes, J.P. Donoghue, V. Thangaraj, R.R. Edelman, S. Warach, Shared neural substrates control-ling hand movements in human motor cortex. Science **268**(5218), 1775–1777 (1995). https://doi.org/10.1126/science.7792606

M.H. Schieber, L.S. Hibbard, How somatotopic is the motor cortex hand area? Science **261**(5120), 489–492 (1993). https://doi.org/10.1126/science.8332915

K.E. Schroeder, Z.T. Irwin, A.J. Bullard, D.E. Thompson, J.N. Bentley, W.C. Stacey, P.G. Patil, C.A. Chestek, Robust tactile sensory responses in finger area of primate motor cortex relevant to prosthetic control. J. Neural. Eng. **14**(4), 46016 (2017). https://doi.org/10.1088/1741-2552/aa7329

S.A. Sheth, C.A. Eckhardt, B.P. Walcott, E.N. Eskandar, M.V. Simon, Factors affecting successful localization of the central sulcus using the somatosensory evoked potential phase reversal technique. Neurosurgery **72**(5), 828–834 (2013). https://doi.org/10.1227/NEU.0b013e3182897447

D. Shin, H. Watanabe, H. Kambara, A. Nambu, T. Isa, Y. Nishimura, Y. Koike, Prediction of muscle activities from electrocorticograms in primary motor cortex of primates. PLoS ONE **7**(10), 1–10 (2012). https://doi.org/10.1371/journal.pone.0047992

M.V. Simon, A.J. Cole, E.C. Chang, B.R. Buchbinder, S.M. Stufflebeam, A. Nozari, A.O. Stemmer-Rachamimov, E.N. Eskandar, An intraoperative multimodal neurophysiologic approach to successful resection of precentral gyrus epileptogenic lesions. Epilepsia **53**(4) (2012). https://doi.org/10.1111/j.1528-1167.2011.03400.x

M.V. Simon, S.A. Sheth, C.A. Eckhardt, R.D. Kilbride, D. Braver, Z. Williams, W. Curry, D. Cahill, E.N. Eskandar, Phase reversal technique decreases cortical stimulation time during motor mapping. J. Clin. Neurosci. **21**(6), 1011–1017 (2014). https://doi.org/10.1016/j.jocn.2013.12.015

D.K. Su, J.G. Ojemann, Electrocorticographic sensorimotor mapping. Clin. Neurophysiol.: Off. J. Int. Fed. Clin. Neurophysiol. **124**(6), 1044–1048 (2013). https://doi.org/10.1016/j.clinph.2013.02.114

H. Tan, A. Pogosyan, K. Ashkan, A.L. Green, T. Aziz, T. Foltynie, P. Limousin, L. Zrinzo, M. Hariz, P. Brown, Decoding gripping force based on local field potentials recorded from subthalamic nucleus in humans. ELife **5**(Nov), 1–24 (2016). https://doi.org/10.7554/eLife.19089

C. Tzagarakis, N.F. Ince, A.C. Leuthold, G. Pellizzer, Beta-band activity during motor planning reflects response uncertainty. J. Neurosci. **30**(34), 11270–11277 (2010). https://doi.org/10.1523/JNEUROSCI.6026-09.2010

S. Waldert, G. Vigneswaran, R. Philipp, R.N. Lemon, A. Kraskov, Modulation of the intracortical LFP during action execution and observation. J. Neurosci. **35**(22), 8451–8461 (2015). https://doi.org/10.1523/JNEUROSCI.5137-14.2015

W. Wang, J.L. Collinger, A.D. Degenhart, E.C. Tyler-Kabara, A.B. Schwartz, D.W. Moran, D.J. Weber, B. Wodlinger, R.K. Vinjamuri, R.C. Ashmore, J.W. Kelly, M.L. Boninger, An electrocorticographic brain interface in an individual with tetraplegia. PLoS ONE **8**(2), e55344 (2013)

B. Wodlinger, J.E. Downey, E.C. Tyler-Kabara, A.B. Schwartz, M.L. Boninger, J.L. Collinger, Ten-dimensional anthropomorphic arm control in a human brain-machine interface: difficulties, solutions, and limitations. J. Neural. Eng. **12**(1), 016011 (2015). https://doi.org/10.1088/1741-2560/12/1/016011

T. Yanagisawa, M. Hirata, Y. Saitoh, T. Goto, H. Kishima, R. Fukuma, H. Yokoi, Y. Kamitani, T. Yoshimine, Real-time control of a prosthetic hand using human electrocorticography signals. J. Neurosurg. **114**(6), 1715–1722 (2011)

T. Yanagisawa, M. Hirata, Y. Saitoh, H. Kishima, K. Matsushita, T. Goto, R. Fukuma, H. Yokoi, Y. Kamitani, T. Yoshimine, Electrocorticographic control of a prosthetic arm in paralyzed patients. Ann. Neurol. **71**(3), 353–361 (2012). https://doi.org/10.1002/ana.22613

# Developing a Closed-Loop Brain-Computer Interface for Treatment of Neuropsychiatric Disorders Using Electrical Brain Stimulation

**Yuxiao Yang, Omid G. Sani, Morgan B. Lee, Heather E. Dawes, Edward F. Chang, and Maryam M. Shanechi**

**Abstract** Neuropsychiatric disorders are a leading cause of disability worldwide. In our research project we are developing brain-computer interface technology to decode mood states that determine appropriate stimulation parameters for real-time therapy.

**Keywords** Depressive disorder · Brain stimulation · Mood states

## 1 Introduction

Neuropsychiatric disorders are a leading cause of disability worldwide, among which depressive disorders are the most debilitating (Whiteford et al. 2013). In the United States alone, every year, an estimated 7.1% of adults (17.3 million) experience at least one major depressive episode ("2018 National Survey of Drug Use and Health

Equal contribution: Yuxiao Yang, Omid G. Sani

Y. Yang · O. G. Sani · M. M. Shanechi (✉)
Ming Hsieh Department of Electrical and Computer Engineering, Viterbi School
of Engineering, University of Southern California, Los Angeles, CA 90089, USA
e-mail: shanechi@usc.edu

M. B. Lee · H. E. Dawes · E. F. Chang
Department of Neurological Surgery, University of California, San Francisco, CA 94122, USA

Weill Institute for Neuroscience,
University of California, San Francisco, CA 94122, USA

Kavli Institute for Fundamental Neuroscience, University of California, San Francisco, CA 94122, USA

M. M. Shanechi
Neuroscience Graduate Program, University of Southern California,
Los Angeles, CA 90089, USA

Department of Biomedical Engineering, Viterbi School of Engineering, University of Southern California, Los Angeles, CA 90089, USA

(NSDUH)" 2018). About 30% of major depressive cases do not respond to current treatments, i.e., are treatment resistant (Rush et al. 2006). Electrical brain stimulation holds promise as a new therapy for treatment-resistant depression. Early seminal studies on open-loop stimulation—which applies stimulation continuously over time without guiding it with neural activity—showed promise in relieving depression symptoms (Mayberg et al. 2005; Lozano et al. 2008; Schlaepfer et al. 2008, 2013; Malone et al. 2009) but had variable efficacy in recent clinical trials (Dougherty et al. 2015; Holtzheimer et al. 2017). Since neuropsychiatric symptoms exhibit inter- and intra-patient variabilities, developing a personalized closed-loop brain-computer interface (BCI) approach could help improve efficacy by precisely tailoring the stimulation therapy to the patient's needs. For each patient, a personalized closed-loop BCI would use the neural activity to decode the relevant mood states related to depression and anxiety symptoms; then these decoded mood states would serve as feedback to determine the appropriate stimulation parameters in real time for a desired therapeutic outcome (Shanechi 2019; Yang et al. 2018). Despite the promise, such a personalized closed-loop BCI is unrealized to date (Shanechi 2019). Indeed, a major obstacle toward realizing such a BCI is the need to build models that can decode mood state variations related to depression and anxiety symptoms over time and to design closed-loop controllers for stimulation (Shanechi 2019; Yang et al. 2018).

The decoding of mood states entails distinct challenges compared with decoding movements in traditional motor BCIs. In motor BCIs, motor intentions can be decoded well from relatively local motor cortical areas (Shanechi 2019). Also, since movements can be measured continuously in time, a large amount of training data can be obtained to train the decoders. In contrast, for neuropsychiatric disorders, distributed multisite brain networks underlie the symptoms (Drevets 2001; Kupfer et al. 2012) which necessitates modelling a high-dimensional multisite neural feature space (Shanechi 2019). Moreover, neuropsychiatric symptoms are difficult to assess, and often need to be measured using questionnaires (Ekkekakis 2013; Widge et al. 2017). This means that only sparse measurements of the symptoms would be available for decoder training. Therefore, decoding mood states poses a challenging modeling problem and had remained elusive (Sani et al. 2018).

## 2 Neural Decoder of Mood State

We developed a new modeling framework to build neural decoders of mood state and demonstrated, for the first time, that mood state variations can be decoded in human epilepsy patients from intracranial human brain activity (Sani et al. 2018). Over multiple days in seven epilepsy subjects, we continuously recorded multisite intracranial EEG (iEEG) and simultaneously collected self-reported mood states that measured depression and anxiety symptoms. We devised a novel modelling framework that can use the sparse mood state measurements to identify a minimal network of brain regions that was sufficient for decoding within the high-dimensional multisite iEEG recordings. The framework trained a mood decoder using the iEEG signals of the identified minimal network. We showed that our modelling framework could significantly decode mood state variations in every individual subject. Across the population, the cross-validated decoding had a correlation coefficient of 0.75 with its true value, leading to an explained variance of 0.57 (Sani et al. 2018). Interestingly, our framework largely selected the limbic regions for decoding, consistent with prior evidence from fMRI studies, which suggest a key role for these regions in depression (Drevets 2001; Kupfer et al. 2012).

Beyond decoding mood variations, another challenge is to design a closed-loop controller that takes the decoded mood as feedback to determine the stimulation parameters to take a pathological brain state toward a healthy state (Shanechi 2019). To design such a closed-loop controller, we need a system identification framework for the effect of stimulation on symptoms (Shanechi 2019; Yang et al. 2018, 2021). To address this challenge, we have developed a control-theoretic system identification framework that uses a novel clinically safe binary-noise modulated stimulation pattern in our prior theoretical work (Yang et al. 2018). We have validated this system identification framework with hardware-in-the-loop simulations (Yang et al. 2018) and more recently in animal models (Yang et al. 2021). We are currently actively investigating how these computational techniques can be leveraged to develop novel models for the effect of brain stimulation in humans in the future. We are also exploring the adaptive tracking of non-stationarity and state-dependency in neural activity over time (Ahmadipour et al. 2020; Yang et al. 2020). Our results to date help pave the way toward future personalized closed-loop BCIs for precisely tailored electrical stimulation therapies for neuropsychiatric disorders, which holds promise to improve the quality of life for millions of patients suffering from treatment-resistant neuropsychiatric conditions. Our work to date has been published in (Shanechi 2019; Yang et al. 2018, 2019, 2020, 2021; Sani et al. 2018; Ahmadipour et al. 2020).

# References

P. Ahmadipour, Y. Yang, E.F. Chang, M.M. Shanechi, Adaptive tracking of human ECoG network dynamics. J. Neural Eng. (2020). https://doi.org/10.1088/1741-2552/abae42

D.D. Dougherty et al., A randomized sham-controlled trial of deep brain stimulation of the ventral capsule/ventral striatum for chronic treatment-resistant depression. Biol. Psychiatry **78**(4), 240–248 (2015). https://doi.org/10.1016/j.biopsych.2014.11.023

W.C. Drevets, Neuroimaging and neuropathological studies of depression: implications for the cognitive-emotional features of mood disorders. Curr. Opin. Neurobiol. **11**(2), 240–249 (2001). https://doi.org/10.1016/S0959-4388(00)00203-8

P. Ekkekakis, *The Measurement of Affect, Mood, and Emotion: A Guide for Health-Behavioral Research* (Cambridge University Press, New York, 2013)

P.E. Holtzheimer et al., Subcallosal cingulate deep brain stimulation for treatment-resistant depression: a multisite, randomised, sham-controlled trial. Lancet Psychiatry **4**(11), 839–849 (2017). https://doi.org/10.1016/S2215-0366(17)30371-1

D.J. Kupfer, E. Frank, M.L. Phillips, Major depressive disorder: new clinical, neurobiological, and treatment perspectives. Lancet **379**(9820), 1045–1055 (2012). https://doi.org/10.1016/S0140-673 6(11)60602-8

A.M. Lozano, H.S. Mayberg, P. Giacobbe, C. Hamani, R.C. Craddock, S.H. Kennedy, Subcallosal cingulate gyrus deep brain stimulation for treatment-resistant depression. Biol. Psychiatry **64**(6), 461–467 (2008). https://doi.org/10.1016/j.biopsych.2008.05.034

D.A. Malone et al., Deep brain stimulation of the ventral capsule/ventral striatum for treatment-resistant depression. Biol. Psychiatry **65**(4), 267–275 (2009). https://doi.org/10.1016/j.biopsych. 2008.08.029

H.S. Mayberg et al., Deep brain stimulation for treatment-resistant depression. Neuron **45**(5), 651–660 (2005). https://doi.org/10.1016/j.neuron.2005.02.014

"2018 National Survey of Drug Use and Health (NSDUH)." (2018). Accessed 11 Apr 2020 [Online]. Available: https://www.samhsa.gov/data/release/2018-national-survey-drug-use-and-health-nsduh-releases

A.J. Rush et al., Acute and longer-term outcomes in depressed outpatients requiring one or several treatment steps: a STAR*D report. Am. J. Psychiatry **163**(11), 1905–1917 (2006). https://doi. org/10.1176/ajp.2006.163.11.1905

O.G. Sani, Y. Yang, M.B. Lee, H.E. Dawes, E.F. Chang, M.M. Shanechi, Mood variations decoded from multi-site intracranial human brain activity. Nat. Biotechnol. **36**, 954 (2018). https://doi.org/ 10.1038/nbt.4200

T.E. Schlaepfer et al., Deep brain stimulation to reward circuitry alleviates anhedonia in refractory major depression. Neuropsychopharmacology **33**(2), 368–377 (2008). https://doi.org/10.1038/sj. npp.1301408

T.E. Schlaepfer, B.H. Bewernick, S. Kayser, B. Mädler, V.A. Coenen, Rapid effects of deep brain stimulation for treatment-resistant major depression. Biol. Psychiatry **73**(12), 1204–1212 (2013). https://doi.org/10.1016/j.biopsych.2013.01.034

M.M. Shanechi, Brain–machine interfaces from motor to mood. Nat. Neurosci. **22**(10), 1554–1564 (2019). https://doi.org/10.1038/s41593-019-0488-y

H.A. Whiteford et al., Global burden of disease attributable to mental and substance use disorders: findings from the Global Burden of Disease Study 2010. Lancet **382**(9904), 1575–1586 (2013). https://doi.org/10.1016/S0140-6736(13)61611-6

A.S. Widge et al., Treating refractory mental illness with closed-loop brain stimulation: progress towards a patient-specific transdiagnostic approach. Exp. Neurol. **287**, 461–472 (2017). https:// doi.org/10.1016/j.expneurol.2016.07.021

Y. Yang, A.T. Connolly, M.M. Shanechi, A control-theoretic system identification framework and a real-time closed-loop clinical simulation testbed for electrical brain stimulation. J. Neural Eng. **15**(6), 066007 (2018). https://doi.org/10.1088/1741-2552/aad1a8

Y. Yang, O.G. Sani, E.F. Chang, M.M. Shanechi, Dynamic network modeling and dimensionality reduction for human ECoG activity. J. Neural Eng. **16**(5), 056014 (2019). https://doi.org/10.1088/ 1741-2552/ab2214

Y. Yang, P. Ahmadipour, M.M. Shanechi, Adaptive latent state modeling of brain network dynamics with real-time learning rate optimization. J. Neural Eng. (2020). https://doi.org/10.1088/1741-2552/abcefd

Y. Yang, S. Qiao, O.G. Sani, J.I. Sedillo, B. Ferrentino, B. Pesaran, M.M. Shanechi, Modelling and prediction of the dynamic responses of large-scale brain networks during direct electrical stimulation. Nat. Biomed. Eng. (2021). https://doi.org/10.1038/s41551-020-00666-w

# Decoding Speech from Dorsal Motor Cortex

Sergey Stavisky

**Abstract** Dr. Stavisky and his team won first place in the BCI Research Award 2019 with their project. Here, we interviewed him to learn more about his project, including the technologies involved and possible clinical applications. Several figures introduce the team and elucidate details about their winning project. The interview concludes with new work and references.

**Keywords** Brain-computer interface · Intracortical · Utah array · Actual speech · Attempted speech · BCI awards

## 1 Introduction

Sergey D. Stavisky from Stanford University, USA and his team detected speech-related neural activity from Utah arrays that had already been placed in the dorsal "arm/hand" area of motor cortex of an intracortical BCI clinical trial participant with tetraplegia. This enabled them to study the motor cortical dynamics during speech production at the unprecedented resolution of populations of single neurons. Their prototype work could identify one of nine spoken syllables with 84.6% accuracy, with 83.5% accuracy among ten words. Their project was:

**Decoding Speech from Intracortical Multielectrode Arrays in Dorsal Motor Cortex**

Sergey D. Stavisky[1], Francis R. Willett[1], Paymon Rezaii[1], Leigh R. Hochberg[2], Krishna V. Shenoy[1,3], Jaimie M. Henderson[1]

[1] Stanford University, USA.

[2] Brown University, Harvard Medical School, Massachusetts General Hospital, Providence VA Medical Center, USA.

[3] Howard Hughes Medical Institute, USA (Fig. 1).

S. Stavisky (✉)
Neurosurgery Department, Stanford University, Stanford, USA
e-mail: sergey.stavisky@stanford.edu

Department of Electrical Engineering, Stanford University, Stanford, USA

C. Guger et al. (eds.), *Brain-Computer Interface Research*,
SpringerBriefs in Electrical and Computer Engineering,
https://doi.org/10.1007/978-3-030-60460-8_10

**Fig. 1** Sergey Stavisky (top row, right) and most of the team that contributed to their award-winning project, including Frank Willett (top row, second from left), Jaimie Henderson (top row, middle), and Krishna Shenoy (top row, second from right). Other current and former members of the Stanford Neural Prosthetics Translational Laboratory in this photo are Eli Stein, Donald Avansino, Guy Wilson, Darrel Deo, and Sharlene Flesher

Sergey and his team won 1st place in the BCI Award 2019 (Fig. 4). We interviewed him about their project and future applications, then edited the interview and added figures and references. In addition to the references that Dr. Stavisky cited in his interview, we provided additional references for readers interested in related publications.

## 2 Interview

**Sergey, you submitted your BCI research "Decoding speech from intracortical multielectrode arrays in dorsal motor cortex" to the BCI Award 2019 and won 1st place. Could you briefly describe what this project was about?**

Sergey:  We recorded directly from inside an area of the brain traditionally thought of as controlling arm and hand movements, and we found that the neural signals there also reflected what the person was speaking. This allowed us to prototype ways to identify what the person was saying, which is a first step to building a BCI for restoring speech.

## What was your goal?

Sergey: This was our team's first foray into speech BCIs, so initially we just wanted to see if we would even see any speech-related neural activity. We already had electrode arrays implanted in the "arm and hand" area of motor cortex of participants in the BrainGate2 BCI pilot clinical trial, and much of our previous work focused on decoding attempted arm movements (for example, see our consortium's 2018 BCI Award submission). While there had been some incredible recent demonstrations of speech decoding using electrocorticography, those studies used signals from much more ventral brain areas than where our arrays were. I didn't have high expectations going in: it was a bit of a "let's take a look and see what we see" if we ask our participants to speak. When we did find speech-related activity, this was both surprising and exciting! From there, we shifted into high gear to try to decode these signals.

## What technologies did you use?

Sergey: We made these recordings using two Blackrock Microsystems 96-electrode arrays (Fig. 2). It's the only intracortical sensor approved for long-term human use, and it's allowed us to learn a great deal about the brain and to demonstrate what is possible using implanted BCI. Looking to the future, I'm excited about the possibility of getting even better neural signals using new and improved implanted neural recording devices. In terms of the

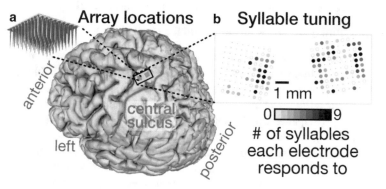

**Fig. 2** Panel **a** shows where the two arrays were placed on the cortex. Panel **b** shows that 73 out 104 functioning electrodes' TC firing rates showed a significance response during speaking, and most responded to speaking multiple syllables

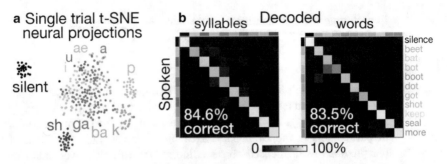

**Fig. 3** Panel **a** shows a two-dimensional representation of individual trials' neural features, demonstrating clustering within syllables and phonetic groupings across syllables. Panel **b** shows classifier confusion matrices: cross-validated prediction accuracies were 84.6% for syllables (10 classes, mean chance accuracy was 10.1% across shuffle controls) and 83.5% for words (11 classes, chance was 9.1%)

algorithms, in this initial work we used conventional statistics techniques, but we're now applying modern deep learning techniques to much larger human speaking datasets (Figs. 2 and 3).

**What kinds of people could benefit from your research?**

Sergey:   Showing that we can decode speech-related activity from implanted electrode arrays is a first step towards building BCIs to restore speech. This could make a tremendous difference for people who have lost their ability to speak, for example due to stroke, traumatic brain injury, ALS, or vocal tract injury. In doing so, we're studying speech production with single neuron resolution, which I hope will lead to fundamental scientific discoveries in addition to the more direct translational applications (Fig. 3).

**Do you think your work has potential for clinical use?**

Sergey:   I absolutely do, but there's a long road to get there. First of all, here we showed that we can identify which of a small number of syllables or words was spoken, in isolation. A clinical speech BCI should be capable of synthesizing a full range of continuous speech. Second, we identified sounds that the participants actually spoke out loud. There are additional challenges in building the map from neural activity to speech if the user isn't able to speak at all. Third, we recorded neural activity using electrodes that have external wires coming out through the scalp. In a clinical system, the sensors need to become fully implanted. Fortunately, there's a lot of work being done on all of these fronts by many groups, including ours (Fig. 4).

**Can you recommend examples of new articles from other groups that address these fronts?**

Sergey:   Yes, there's really been a hot streak in the speech BCI subfield—exciting times! Just in the past year, there have been multiple very impressive

**Fig. 4** 3D reconstruction of the two 100-electrode Utah arrays in the participant's dorsal precentral gyrus (the so-called 'hand knob' area of motor cortex). An example of the simultaneously recorded speech audio waveform (*blue*) and neuron action potentials (*yellow* and *pink* ticks) are shown

recent studies for decoding continuous speech using ECoG recordings. Gopala Anumanchipalli and Josh Chartier led one such study reconstructing the produced speech movements and sounds during spoken sentences (Anumanchipalli et al. 2019), while Christian Herff and Miguel Angrick led two other studies using a mix of unit selection and deep learning methods, respectively, to reconstruct spoken words (Herff et al. 2019; Angrick et al. 2019). There was also an innovative study by Joseph Makin which used end-to-end machine translation methods to decode spoken sentences, and this work was a finalist for the 2020 BCI Award (Makin et al. 2020).

Another approach is to decode from auditory areas, with the idea being that one might decode the imagined "inner voice" of patients unable to speak. To that end, a recent study from Hassan Akbari and colleagues (Akbari et al. 2019) showed reconstruction of heard speech from human auditory cortex, while a team led by Christopher Heelan and Jihun Lee reconstructed heard speech from monkey auditory cortex using Utah arrays (Heelan et al. 2019).

For moving speech BCIs towards closed-loop systems, a project led by David Moses (Moses et al. 2019) showed a real-time question-and-answer decoder (restricted to a small number of possible response). Other groups are also exploring intracortical approaches in people. For example, one way to get intracortical data while people speak is stereotactic-EEG, and there's a nice recent review by Herff, Krusienski, and Kubben summarizing their (and others') recent work in that domain (Herff et al. 2020).

**Do you have any new work since the award ceremony related to your project?**

Sergey:   Yes, we've recently published two additional studies that built upon this
          discovery of speech-related activity in dorsal motor cortex.

The first of these new studies addressed the concern that this speech-related
activity would be a "nuisance variable" during arm/hand BCI use. For example,
imagine if a BCI user speaks while trying to control an arm prosthesis (or a computer
cursor): would the accompanying modulation of the same neurons that also are linked
to moving the arm inadvertently affect the BCI output and interfere with achieving
the intended reach goal? To answer this, we conducted a follow-up study in which our
participant spoke while simultaneously using a cursor BCI. Fortunately, we found
that the interference scenario just was not the case: when the participant spoke while
simultaneously attempting arm movements, speech-related activity in dorsal motor
cortex was attenuated and did not affect the decoder. We also asked the participant
whether it was difficult to talk while using the cursor BCI, and he said it was not. This
is encouraging as it shows that the "biomimetic" motor BCIs we are developing—
meaning, where the person attempts to move their arm as they normally would—are
intuitive and not overly cognitively demanding. This study is now published in the
Journal of Neural Engineering (Stavisky et al. 2020).

In the second study (Wilson et al. 2020), my colleague Guy Wilson and I extended
our dorsal motor cortex speech decoding results to cover more than the 9 or 10-
class prediction from our 2019 BCI Award project (and associated *eLife* paper).
Specifically, we trained decoders to discriminate amongst a comprehensive speech
basis set of 39 English phonemes, or to directly reconstruct speech audio using
the 'brain-to-text' approach from the aforementioned Herff et al. (2019) study. We
achieved up to 39% accuracy for the phoneme prediction and r = 0.52 correlation
between true and reconstructed audio, which is competitive with prior ECoG work
despite our recording from what is almost certainly a suboptimal area of the brain.
These results give me a lot of confidence that we can do even better with intracortical
measurements from more ventral speech areas of cortex.

**What was it like to win the BCI Award 2019?**

Sergey:   It was fantastic news! The whole team was delighted. There's so much
          effort over many, many months that goes into BCI research (for example,
          we started this project in Autumn 2017), so it's really nice to have awards
          like this that come as a pleasant surprise and recognize the work.

**How was your research funded?**

Sergey:   We are very grateful for support from a number of funders, including
          both U.S. federal sources and private foundations, which are listed in the
          Acknowledgements section, which made this research possible.

**Acknowledgements** This project was supported by an ALS Association Milton Safenowitz Postdoctoral Fellowship, A. P. Giannini Foundation Postdoctoral Fellowship, Wu Tsai Neurosciences
Institute Interdisciplinary Scholar Award, and Burroughs Wellcome Fund Career Award at the Scientific Interface; NSF Graduate Research Fellowship DGE—1656518 and Regina Casper Stanford

Graduate Fellowship; Larry and Pamela Garlick, Samuel and Betsy Reeves; NIDCD R01DC014034; Office of Research and Development, Rehabilitation R and D Service, Department of Veterans Affairs N9288C, A2295R, B6453R, Executive Committee on Research of Massachusetts General Hospital, NIDCD R01DC009899; NINDS 5U01NS098968-02; and Howard Hughes Medical Institute.

# References

H. Akbari, B. Khalighinejad, J.L. Herrero, A.D. Mehta, N. Mesgarani, Towards reconstructing intelligible speech from the human auditory cortex. Sci. Rep. **9**, 874 (2019)

M. Angrick, C. Herff, E. Mugler, M.C. Tate, M.W. Slutzky, D.J. Krusienski, T. Schultz, Speech synthesis from ECoG using densely connected 3D convolutional neural networks J. Neural Eng. **16**, 036019 (2019)

G.K. Anumanchipalli, J. Chartier, E.F. Chang, Speech synthesis from neural decoding of spoken sentences. Nature **568**, 493–498 (2019)

C. Heelan, J. Lee, R. O'Shea, L. Lynch, D.M. Brandman, W. Truccolo, A.V. Nurmikko, Decoding speech from spike-based neural population recordings in secondary auditory cortex of non-human primates Commun. Biol. **2**, 466 (2019)

C. Herff, L. Diener, M. Angrick, E. Mugler, M.C. Tate, M.A. Goldrick, D.J. Krusienski, M.W. Slutzky, T. Schultz, Generating natural, intelligible speech from brain activity in motor, premotor, and inferior frontal cortices. Front. Neurosci. **13**, 1–11 (2019)

C. Herff, D.J. Krusienski, P. Kubben, The potential of stereotactic-EEG for brain-computer interfaces: current progress and future directions. Front. Neurosci. **14**, 1–8 (2020)

J.G. Makin, D.A. Moses, E.F. Chang, Machine translation of cortical activity to text with an encoder–decoder framework. Nat. Neurosci. **23**, 575–582 (2020)

D.A. Moses, M.K. Leonard, J.G. Makin, E.F. Chang, Real-time decoding of question-and-answer speech dialogue using human cortical activity. Nat. Commun. **10**, 3096 (2019)

S.D. Stavisky, F.R. Willett, D.T. Avansino, L.R. Hochberg, K.V. Shenoy, J.M. Henderson, Speech-related dorsal motor cortex activity does not interfere with iBCI cursor control. J. Neural Eng. **17**(1), 016049 (2020)

G.H. Wilson, S.D. Stavisky, F.R. Willett, D.T. Avansino, J.N. Kelemen, L.R. Hochberg, J.M. Henderson, S. Druckman, K.V. Shenoy, Decoding spoken English phonemes from intracortical electrode arrays in dorsal precentral gyrus. J. Neural Eng., 9 Nov 2020. https://www.biorxiv.org/content/10.1101/2020.06.30.180935v1.abstract

# Training with BCI-Based Neurofeedback for Quitting Smoking

**Junjie Bu**

**Abstract** Dr. Bu and his team won second place in the BCI Research Award 2019 with their project that used neurofeedback training to help people quit smoking. Chapter "BCI-based Neurofeedback Training for Quitting Smoking" of this book describes this project in detail, while this chapter presents an interview with Dr. Bu. This interview addresses non-technical issues such as how the idea for their project developed, the importance of working with people from different backgrounds, advice for people new to BCI research, recent work since the BCI Research Award, and how they plan to extend their work in the next several years.

**Keywords** Brain-computer interface · Cognition-guided neurofeedback · Nicotine addiction · Smoking cue reactivity · BCI Research Awards

## 1 Introduction

Dr. Junjie Bu studied at the University of Science & Technology of China and is currently with Anhui Medical University. His team developed a closed-loop neurofeedback training using a brain-computer interface to help smokers quit. Their approach reduced cigarette craving and smoking behaviour and is a promising BCI-based tool for treating addiction. Their project was:

**BCI-based Neurofeedback Training for Quitting Smoking**

Junjie Bu[1], Kymberly D. Young[2], Wei Hong[1], Ru Ma[1], Hongwen Song[5], Ying Wang[1], Wei Zhang[1],

Michelle Hampson[3], Talma Hendler[4], Xiaochu Zhang[1,5].

[1]Hefei National Laboratory for Physical Sciences at the Microscale and School of Life Sciences, University of Science & Technology of China, Hefei, China.

---

J. Bu (✉)
School of Biomedical Engineering, Research and Engineering Center of Biomedical Material, Anhui Medical University, Hefei, China
e-mail: bujunjie@ahmu.edu.cn

School of Life Sciences, University of Science & Technology of China, Hefei, China

[2]Department of Psychiatry, University of Pittsburgh School of Medicine, Pittsburgh, USA.

[3]Department of Radiology and Biomedical Imaging, Yale School of Medicine, New Haven, CT, USA.

[4]Functional Brain Center, Tel-Aviv University, Tel-Aviv, Israel.

[5]School of Humanities & Social Science, University of Science & Technology of China, Hefei, China.

This project won 2nd place at the BCI Award 2019. Dr. Bu and the senior author, Prof. Zhang, describe their project in chapter "BCI-based Neurofeedback Training for Quitting Smoking" of this book. That chapter includes background information, figures describing the approach and results, statistical analyses, references, and other details. This interview complements that chapter with an easier overview of their project, along with how their project developed, new work, and future directions.

This interview chapter is based on an interview that Dr. Guger conducted with Dr. Bu. The book editors then worked with Dr. Bu via email to edit the interview and add additional text, figures, acknowledgments, and references.

## 2  Interview

**Junjie, you submitted your BCI research "BCI-based neurofeedback training for quitting smoking" to the BCI Award 2019 and won 2nd place. Could you briefly describe what this project was about?**

**Junjie:** Yes! In this project, we developed a novel cognition-guided neurofeedback system and tested its therapeutic efficacy on nicotine addiction using a randomized clinical trial. Using this neurofeedback, smokers were trained to de-activate their EEG activity patterns related to smoking cue reactivity. We found that this neurofeedback produced short-term and long-term effects on cigarette craving and smoking behaviour. In particular, the rate of smoking decreased as much as 38.2% during the 4-month follow-up period after only two sessions of this neurofeedback training.

**What was your goal?**

**Junjie:** We have been applying the cognition-guided neurofeedback approach for treating methamphetamine addiction and alcohol addiction. As we know, they are more severe than nicotine addiction. However, there are few effective treatments for them. We hope that our neurofeedback could help patients reduce the symptoms of addiction and change their lives. We are also working to improve training for different patients by optimizing different aspects of our neurofeedback system, such as the cognitive task, machine learning algorithm, training sessions and so on.

**What technologies did you use?**

**Junjie:** Our cognition-guided neurofeedback consisted of two parts. First, we trained a personalized classifier to distinguish the EEG activity patterns corresponding to smoking and neutral cue reactivity using the specific cognitive task (smoking cue reactivity task). Next, during neurofeedback training, participants were asked to repeatedly and continuously deactivate their real-time EEG activity patterns of smoking cue reactivity calculated using a previously constructed classifier.

**How did you develop the idea to work on that project?**

**Junjie:** My undergraduate background is biomedical engineering. I like cognitive neuroscience and I wanted to apply my knowledge of brain science to make a change, especially to help people. Then, I met my PhD advisor, Prof. Xiaochu Zhang. He is a cognitive psychologist. We used our strengths together to start the neurofeedback project.

**Which disciplines are involved?**

**Junjie:** First, we designed a cognitive task based on cognitive psychology. Then, we recorded and analysed the EEG data based on biomedical engineering. Third, we applied a machine learning algorithm for brain pattern recognition based on computer science. Finally, we designed a randomized clinical trial to test the effects based on psychiatry medicine. So it is really interdisciplinary.

**How effective was your approach?**

**Junjie:** Well, after two visits of the neurofeedback training, smokers showed a significant decrease in cigarette craving and craving-related P300 amplitudes. The rates of cigarettes smoked per day at 1 week, 1 month and 4 months follow-up decreased 30.6%, 38.2%, and 27.4% relative to baseline.

**How did it feel to be one of the BCI Award winners?**

**Junjie:** It is a great honor to win second place. My good friend Haohao came with me to take part in the ceremony. The ceremony was great (Fig. 1). When I heard my name, it was really surprising. At the ceremony, I was very grateful and very thankful to g.tec, the BCI Award committee, my PhD advisor Prof. Xiaochu Zhang, my current affiliation with Anhui Medical University (Fig. 2) and my family.

**In Chapter "BCI-based Neurofeedback Training for Quitting Smoking" of this book, you and Prof. Zhang said that this approach merits further testing. Have you conducted any new research to follow up on neurofeedback for quitting smoking?**

**Junjie:** Yes. We published some papers recently with relevant research. Some of these papers detail work based on our project submission (Bu et al. 2019a; Cheng et al. 2020). We also have related new work involving hypnosis to help people quit smoking

**Fig. 1** This image shows five people onstage during the Awards Ceremony. From left to right: Emcee Dr. Krausz, Interviewee Dr. Bu, Chair of the Jury Dr. Tangermann, Jury Member Dr. Wreissneggar, and Emcee Dr. Allison

**Fig. 2** Anhui Medical University is one of the oldest universities in the province of Anhui and spans over 850,000 square meters

and advanced EEG imaging to study effects such as reduced cravings (Bu et al. 2019b; Li et al. 2019a, 2019b). In addition, we plan to combine our neurofeedback with other methods (e.g., hypnosis and tDCS) in order to help them quit smoking better.

**What are the remaining steps before this approach is ready for clinical practice?**

**Junjie:** Although we did conduct a study with many patients, more work is needed before this approach is safe for the public. For example, we need to know more about possible risks or side effects, especially for some groups of patients at higher risk. We might make the approach more effective with more research. Clinical device approval is also necessary.

**Do you have advice for undergraduate or graduate students who want to work on BCIs?**

**Junjie:** You'd better communicate with people who have different backgrounds, including neuroscience, cognitive psychology, psychiatry medicine and so on. Don't restrict your thoughts based on what you have learned and what you previously thought.

**How was your research funded?**

**Junjie:** We appreciate support from the National Key Basic Research Program (2016YFA0400900 and 2018YFC0831101), The National Natural Science Foundation of China (31771221, 31471071, 61773360, 71874170, and 32000750), Anhui Provincial Natural Science Foundation (2008085QH369), China Postdoctoral Science Foundation (2019TQ0312 and 2019M662203), and School Foundation of Anhui Medical University (2020xkjT020, 2019xkj016 and XJ201907).

# References

J. Bu, K.D. Young, W. Hong, R. Ma, H. Song, Y. Wang, W. Zhang, M. Hampson, T. Hendler, X. Zhang, Effect of deactivation of activity patterns related to smoking cue reactivity on nicotine addiction. Brain **142**(6), 1827–1841 (2019a)

J. Bu, R. Ma, C. Fan, S. Sun, Y. Cheng, Y. Piao, P. Zhang, C. Liu, X. Zhang, Low-Theta electroencephalography coherence predicts cigarette craving in nicotine addiction. Front. Psychiatry **10**, 296 (2019b)

Y. Cheng, J. Bu, N. Li, J. Li, H. Gou, S. Sun, C. Liu, Z. Jin, C. He, C. Fan, C. Liu, Dysfunctional resting-state EEG microstate correlated with the severity of cigarette exposure in nicotine addiction. Inf. Sci. **63**(170107), 1–170107 (2020)

X. Li, L. Chen, R. Ma, H. Wang, L. Wan, Y. Wang, J. Bu, W. Hong, W. Lv, S. Vollstädt-Klein, Y. Yang, The top-down regulation from the prefrontal cortex to insula via hypnotic aversion suggestions reduces smoking craving. Hum. Brain Mapp. **40**(6), 1718–1728 (2019a)

X. Li, L. Chen, R. Ma, H. Wang, L. Wan, J. Bu, W. Hong, W. Lv, Y. Yang, H. Rao, X. Zhang, The neural mechanisms of immediate and follow-up of the treatment effect of hypnosis on smoking craving. Brain Imaging Behav. 1–11 (2019b)

# Closed-Loop BCI for the Treatment of Neuropsychiatric Disorders

**Omid G. Sani, Yuxiao Yang, and Maryam M. Shanechi**

**Abstract** Prof. Shanechi and team submitted a project about developing neuro-feedback to help patients with neuropsychiatric disorders such as depression. Their project won third place in the BCI Award 2019. In this chapter, we interviewed Prof. Shanechi and two other members of her team about their project. We asked about why their research is important for patients, how their system operates and might be adapted to real-world use, remaining challenges, and new work from their group and other groups.

**Keywords** Brain-computer interface · Intracranial electroencephalography · Depression · Neurofeedback · BCI Awards

## 1 Introduction

Many patients with depression find therapy, medication, cognitive and behavioral changes, and/or other treatments helpful. However, these methods are not effective for about a quarter of patients with major depression, and thus tens of millions of people worldwide need new ways to treat their depression. The 3rd winner in the 2019 BCI Award presented their work toward realizing a BCI system that could provide an alternative therapy to help patients with major depression for whom other treatments are not effective. Chapter "Developing a Closed-loop Brain-Computer Interface for Treatment of Neuropsychiatric Disorders Using Electrical Brain Stimulation" of this book contains details about their project, while this chapter presents an interview with some members of the winning team. Their project, including the team members behind the project and their affiliations, was:

O. G. Sani (✉) · Y. Yang · M. M. Shanechi
Ming Hsieh Department of Electrical and Computer Engineering, Viterbi School of Engineering, University of Southern California, Los Angeles, CA, USA

M. M. Shanechi
Neuroscience Graduate Program, University of Southern California, Los Angeles, CA, USA

### Developing a Closed-loop Brain-computer Interface for Treatment of Neuropsychiatric Disorders Using Electrical Brain Stimulation

Yuxiao Yang[1,†], Omid G. Sani[1,†], Morgan B. Lee[2,3,4], Heather E. Dawes[2,3,4], Edward F. Chang[2,3,4], Maryam M. Shanechi[1,5].

[1]Ming Hsieh Department of Electrical and Computer Engineering, Viterbi School of Engineering, University of Southern California, USA.

[2]Department of Neurological Surgery, University of California, USA.

[3]Weill Institute for Neuroscience, University of California, San Francisco, USA.

[4]Kavli Institute for Fundamental Neuroscience, University of California, San Francisco, USA.

[5]Neuroscience Graduate Program, University of Southern California, USA.

†Equal author contribution.

This interview chapter is unique in that we had the opportunity to interview not one but three members of the team behind a winning project. Maryam Shanechi, and her PhD students Omid Sani, and Yuxiao Yang answered questions about their project, which won third place in the BCI Award. Their project entailed a collaboration between Maryam Shanechi's Lab at the University of Southern California (USC) and Edward Chang's Lab at the University of California at San Francisco (UCSF):

Figure 1 shows the three interviewees. Dr. Guger initially interviewed them, and then we worked together to develop this chapter, including introductory text, new questions and answers, figures, and references. Readers might also be interested in two chapters in this book from Dr. Bu and colleagues, who won second place this year. These chapters feature a project description (Chapter "BCI-based Neurofeedback Training for Quitting Smoking") and an interview with Dr. Bu about their team's work involving neurofeedback with implanted BCIs to help people quit smoking (Chapter "Decoding Speech from Dorsal Motor Cortex"). Chapter "Neurofeedback of Scalp Bi-Hemispheric EEG Sensorimotor Rhythm Guides Hemispheric Activation of Sensorimotor Cortex in the Targeted Hemisphere" from Hayashi and colleagues describes an EEG-based neurofeedback system to support sensorimotor feedback, which could support neural rehabilitation. Hence, both invasive and non-invasive

**Fig. 1** From left to right: Yuxiao Yang, Maryam Shanechi, and Omid Sani

BCIs that centered on neurofeedback to help different patient groups were prominent among this year's nominees and winners.

## 2  Interview

**Maryam, Yuxiao, and Omid, you submitted your BCI research project titled "Developing a closed-loop brain-computer interface for treatment of neuropsychiatric disorders using electrical brain stimulation" to the BCI Award 2019 and won 3rd place. Could you briefly describe what this project was about?**

**Maryam:** Neuropsychiatric disorders such as major depression are a leading cause of disability worldwide. Currently, about 20–30% of major depression patients are not responsive to any available treatments. This is about 5 million people in the US alone who could greatly benefit from a novel treatment. Motivated by the view that neuropsychiatric disorders are signs of abnormal brain network activity, our goal is to provide a new treatment in the form of a novel closed-loop BCI that aims to normalize these brain activity patterns using electrical stimulation.

To do so, we take a principled engineering approach to designing this BCI. Our aim is to design a BCI that can decide in real time how to stimulate the brain guided by novel decoders that track a patient's mood state that is related to their depression and anxiety symptoms from their brain activity (Shanechi 2019). The goal of the BCI is to alleviate these symptoms by applying the stimulation at the right time, with the right amount, and in a manner that is precisely tailored to the patient's needs. Such a BCI is unrealized to date. In this project, we describe our recent progress toward realizing such a BCI centered around developing decoders and system identification methods for stimulation.

**What technologies did you use?**

**Yuxiao:** We use tools from machine learning and control theory to develop novel mathematical models that explain how a patient's mood symptoms are represented in their brain activity (Sani et al. 2018; Yang et al. 2019). We also use these computational tools to understand how stimulation might normalize the abnormal brain activity patterns (Yang et al. 2018; Shanechi 2019; Yang et al. 2021). For example, to decode mood from brain activity, in collaboration with Edward Chang's team at UCSF, we recorded multi-site intracranial electroencephalography (iEEG) signals from seven epilepsy patients and concurrently measured their mood using a validated self-report questionnaire (Sani et al. 2018). Then, we developed a novel model that can match each patient's iEEG signal with the mood report.

Based on this model, we built a decoder that automatically estimates real-time mood variations from the patient's brain activity. Our future goal is to develop a controller that optimally adjusts the amount of electrical stimulation in real-time to alleviate mood symptoms, thus realizing a closed-loop brain stimulation system (Shanechi 2019).

**How would therapy for the treatment of neuropsychiatric disorders work using your invention?**

**Omid:** We envision that, in the future, our BCI will be able to help provide a precisely-tailored alternative therapy for treatment-resistant major depression. The BCI will tailor the delivery of electrical stimulation to each individual patient's need by monitoring their mood symptoms in real time based on their brain activity. The future goal will be to alleviate symptoms by applying electrical stimulation only when needed, and only with the minimal optimal amount needed.

From the patient's point of view, the treatment will be similar to how standard deep brain stimulation (DBS) systems are currently implanted to treat Parkinson's disease for thousands of patients each year. But, of course, the algorithmic technology within the implant will need to enable closed-loop stimulation and address the distinct challenges for neuropsychiatric disorders. The vision is that after implantation, the patient will be able to go back to their normal life while the device will keep providing the right amount of therapy. Of course, we still have to do much more work to realize this vision in the future. We hope that the progress we have made so far will help facilitate such a BCI.

**Do you work together with other institutions?**

**Maryam:** Yes, the mood decoding work was a close collaboration with Edward Chang's team at UCSF. We believe collaborations are key to success in this truly interdisciplinary domain of BCI design.

**How long will it take to have your technology available?**

**Omid:** We are only at the early stages of moving toward making closed-loop stimulation treatments for neuropsychiatric disorders clinically feasible. So far, we have provided the first demonstration that mood symptoms related to depression and anxiety can be decoded from brain activity. Our next step is to develop models that can decide how to change the stimulation to normalize brain activity patterns underlying these disease symptoms, and to test these models in animal and human experiments. We have recently made progress toward validating such stimulation models in animal models (Yang et al. 2021). We are also exploring the adaptive tracking of state-dependency and non-stationarity in neural signals over time (Ahmadipour et al. 2020; Yang et al. 2020). Finally, we will need to build a real-time optimal controller that can deliver the stimulation at the right time and with the right amount. Once we develop the technology and rigorously validate it in animal models and human experiments, the technology needs to be tested in carefully designed clinical trials to assess efficacy and longevity for a larger group of patients and to validate safety.

**How did it feel to be one of the winners of the BCI Award 2019?**

**Yuxiao:** This is a great recognition of our work toward developing novel BCI technologies that can provide alternative new therapies for millions of patients with treatment-resistant neuropsychiatric disorders. We sincerely thank the jury for selecting our work as one of the winners.

**Can you tell us a little more about your backgrounds?**

**Omid:** I am currently a postdoctoral scholar at USC, where I work with Prof. Shanechi on developing new techniques for studying the brain and building BCIs for treatment of neural disorders. I got the opportunity to work on this project as a part my PhD in electrical engineering in Prof. Shanechi's lab.

**Yuxiao:** I am currently an Assistant Professor at the Department of Electrical and Computer Engineering (ECE) in University of Central Florida (UCF). I worked on this project during my PhD in electrical engineering in Prof. Shanechi's lab at USC.

**Maryam:** I am assistant Professor and Viterbi Early Career Chair at USC Viterbi School of Engineering. I received my PhD at MIT in electrical engineering and computer sciences. My lab develops neurotechnology and studies the brain through decoding and control of neural dynamics.

**Acknowledgements** This project has been partly supported by the following: Defense Advanced Research Projects Agency (DARPA) under Cooperative Agreement Number W911NF-14-2-0043 (to M.M.S. and E.F.C.), issued by the Army Research Office contracting office in support of DARPA's SUBNETS program. Army Research Office (ARO) under contract W911NF-16-1-0368 (to M.M.S) as part of the collaboration between US DOD, UK MOD and UK Engineering and Physical Research Council (EPSRC) under the Multidisciplinary University Research Initiative (MURI). The views, opinions and/or findings expressed are those of the author(s) and should not be interpreted as representing the official views or policies of the Department of Defense or the US Government.

# References

P. Ahmadipour, Y. Yang, E.F. Chang, M.M. Shanechi, Adaptive tracking of human ECoG network dynamics. J. Neural Eng. (2020). https://doi.org/10.1088/1741-2552/abae42

O.G. Sani, Y. Yang, M.B. Lee, H.E. Dawes, E.F. Chang, M.M. Shanechi, Mood variations decoded from multi-site intracranial human brain activity. Nat. Biotechnol. **36**, 954 (2018). https://doi.org/10.1038/nbt.4200

M.M. Shanechi, Brain–machine interfaces from motor to mood. Nat. Neurosci. **22**(10), 1554–1564 (2019). https://doi.org/10.1038/s41593-019-0488-y

Y. Yang, A.T. Connolly, M.M. Shanechi, A control-theoretic system identification framework and a real-time closed-loop clinical simulation testbed for electrical brain stimulation. J. Neural Eng. **15**(6), 066007 (2018). https://doi.org/10.1088/1741-2552/aad1a8

Y. Yang, O.G. Sani, E.F. Chang, M.M. Shanechi, Dynamic network modeling and dimensionality reduction for human ECoG activity. J. Neural Eng. **16**(5), 056014 (2019). https://doi.org/10.1088/1741-2552/ab2214

Y. Yang, P. Ahmadipour, M.M. Shanechi, Adaptive latent state modeling of brain network dynamics with real-time learning rate optimization. J. Neural Eng. (2020). https://doi.org/10.1088/1741-2552/abcefd

Y. Yang, S. Qiao, O.G. Sani, J.I. Sedillo, B. Ferrentino, B. Pesaran, M.M. Shanechi, Modelling and prediction of the dynamic responses of large-scale brain networks during direct electrical stimulation. Nat. Biomed. Eng. (2021). https://doi.org/10.1038/s41551-020-00666-w

# The Stentrode™ Neural Interface System

**Nicholas Opie**

**Abstract** Dr. Opie and his team developed the Stentrode™ Neural Interface System, which records brain activity from inside the brain without requiring open-brain surgery. This approach could have many medical applications. Although it has been validated successfully so far, clinical trials, device approval, and more research will be necessary before widespread application in people. This chapter presents an interview with Dr. Opie about the project that his team submitted to the BCI Award 2019, which was nominated for an award. Readers can learn about their team's work developing the Stentrode™, how it works, the research they conducted within the project submission, new work, and future directions.

**Keywords** Brain-Computer Interface · Stentrode™ Neural Interface System · Paralysis · Electrode array · BCI Awards

## 1 Introduction

The Stentrode™ Neural Interface System is a minimally invasive, wireless BCI technology that records brain signals from electrodes positioned within cerebral blood vessels. This approach overcomes limitations of existing BCIs, including the need for open-brain surgeries, degradation of signals due to inflammation, tissue reactions to penetrating microelectrodes, unilateral neural signal recording and required expertise for system use. The Stentrode™ Neural Interface System can sense bilateral brain signals from information-rich cortical areas without penetrating the skull or dura and perform processing tasks to achieve BCI control using a custom-built software platform requiring no expert knowledge for use. Feasibility of this approach is supported by promising results from large-animal studies, which demonstrate the potential for the Stentrode™ Neural Interface System to become the first minimally invasive

N. Opie (✉)
Synchron Inc., Melbourne, VIC, Australia
e-mail: n.opie@synchronmed.com

Vascular Bionics Laboratory, Department of Medicine, Melbourne, VIC, Australia

© The Author(s), under exclusive license to Springer Nature Switzerland AG 2021
C. Guger et al. (eds.), *Brain-Computer Interface Research*,
SpringerBriefs in Electrical and Computer Engineering,
https://doi.org/10.1007/978-3-030-60460-8_13

**Fig. 1** This team photo shows the people behind the Stentrode™ work presented in this book chapter

Brain-Computer Interface designed for everyday use. First-in-human clinical trials of the Stentrode™ Neural Interface System are currently underway.

Figure 1 shows Nicholas Opie and his colleagues. They submitted a project involving the Stentrode™ Neural Interface System to the 2019 BCI Award, and their project was nominated as one of the best projects that year. We interviewed Dr. Opie about this project, including some questions about how it addressed the different scoring criteria that the jurors used to evaluate the projects described in the introduction chapter of this book. Then, we (the book editors) worked with Dr. Opie via email to develop the interview into this book chapter. Their project was:

**Stentrode™ Neural Interface System: Minimally-invasive Brain-Computer Interface Designed for Everyday Use**.

Peter Yoo[1], Nicholas Opie[1], Thomas Oxley[1], Stephen Ronayne[1], Gil Rind[1], and Amos Meltzer[1].

[1]Synchron Inc., Australia.

## 2  Interview

### What was the goal of your project?

**Nicholas:** Over the last few years, it has been our goal to design, develop and demonstrate the safety and efficacy of a novel endovascular neural interface that has the potential to restore communication and independence to people with paralysis (Fig. 2). Our technology is designed to record neural information from the motor cortex without subjecting patients to risky surgery to remove the skull to access the brain—surgery that is currently required by all existing implantable brain-machine interfaces.

### How did you approach this goal?

**Nicholas:** We believe that patient safety is paramount. Consequently, we approached this challenge by identifying methods to access the brain from within the skull, without skull removal. To mitigate the risks associated with open-brain surgery, we developed a method of accessing the brain and motor cortex via blood vessels, using angiographic surgical techniques commonly practiced to remove cortical blood clots. We have built a human-grade device, validated preclinical testing and have implanted Australian participants with large success (Fig. 2).

### What technology did you use and why?

**Nicholas:** We have fabricated a monolithic stent electrode array (the Stentrode™) for permanent implantation within a blood vessel. This device is delivered to the motor cortex through a small catheter. When in the desired location, the catheter is removed, and the Stentrode™ self-expands to conform to the curvature of the vessel, placing the electrodes against the vessel wall in close proximity to neural tissue. Electrodes pick up neural signals and relay these to a transcutaneous telemetry unit placed in the chest. In turn, the signals are wirelessly transmitted to an external processing

**Fig. 2** These two images illustrate the Stentrode™ system. The left panel depicts the self-expanding electrode array being deployed from within a catheter inside the superior sagittal sinus over the motor cortex and the right panel shows a schematic of a patient using the wireless system to operate a personal computer

unit that interprets and decodes the information, translating it into commands that can be used to control assistive technology such as computers, wheelchairs and exoskeletons.

### Does the approach work online/in real-time?

**Nicholas:** Yes. The system is designed for real-time decoding and has been bench tested with simulated and pseudo-signals for real-time feature extraction, classification, translation into digital commands and interfacing with devices and applications.

### Does the project include any novel hardware or software developments, or a novel application of the BCI?

**Nicholas:** The Stentrode™ Neural Interface is the first fully-implanted wireless endovascular BCI capable of recording high-fidelity broadband neural signals without the need for invasive open brain surgeries or percutaneous connections. Unlike previous research systems, the Stentrode™ Neural Interface System incorporates deliberate design features, such as fully implanted battery-less ITU, a wireless ETU for device powering and data transfer and a custom-built software that enables everyday use of BCIs for digital device control without specialist knowledge to maximize the effective use by real-life users.

### Is there any new methodological approach used compared to earlier projects?

**Nicholas:** The Stentrode™ overcomes limitations of the existing BCI technology, addressing both issues of safety and efficacy. We performed numerous proof-of-concept studies to investigate demonstrate the safety and efficacy of our technology, which have been published in peer-reviewed journals, including Nature Biotechnology (Oxley et al. 2016; Opie et al. 2016, 2017, 2018). Our website[1] has a full publication list. These studies demonstrated that our device can be implanted in between the two hemispheric motor cortices within a blood vessel in a large animal model, that Stentrode™ can record contralateral movement-related neural information chronically. Using surgical procedures comparable to those practiced for vascular clot removal, we have optimised our technology and demonstrated the capacity to deliver device in an ovine model. Further, we have empirically demonstrated that our technology can record high-frequency bandwidth signal, similar to conventional surface electrode arrays. Further, we have demonstrated preliminary safety of our technology using high-resolution x-ray imaging and histology data of 20 animals implanted for durations of up to 190 days (Opie et al. 2017). We observed that after eight days, endothelial growth covered the stent and mounted electrodes, significantly reducing the potential for clotting and thrombosis caused by exposure of bare metal. No evidence of vascular occlusion of the implanted sinus was observed in any of the animals throughout the 190-day timeframe. We have also conducted rigorous preclinical testing of the Stentrode™ under the guidance of the FDA and local IRBs

---

[1]https://www.synchronmed.com/research/.

to ensure the device meets functional and safety requirements required for implantation in humans. We have also developed a custom-built software that focuses on user-experience and robustness that can perform an exhaustive list of tasks required for real-time BCI control of digital devices. We have commenced the first-in-human clinical trial of the Stentrode™ Neural Interface system, with the primary outcome of safety and secondary outcome of efficacy.

**What were the results?**

**Nicholas**: Our preclinical studies have demonstrated a 95% endothlelialisation at 45 days, indicative that the Stentrode is quickly incorporated into the blood vessel. Not only does this help to maintain vessel patency, but it also anchors the electrodes in place. Further, we have observed that signal quality increases following endothelialisation. Through substantial FDA-required mechanical and biocompatibility testing, we have also observed that the Stentrode is safe for human implantation, and can remain safe and function over a chronic duration.

**How important are the results for clinical environments?**

**Nicholas:** The preclinical testing, incorporating mechanical, biocompatibility and safety assessment of the technology in conjunction with Australian first-in-human evidence are critical for our submission to the FDA for approval under IDE to commence an early feasibility study in the USA. We have demonstrated and documented the manufacturing capabilities of our team, and are poised to continue our journey towards commercial approval.

**Are your results applicable in clinical/real-life environments?**

**Nicholas:** The results we have obtained provide us and the paralysed community with hope that a treatment to their condition is on the horizon. While we are still a number of years away from achieving FDA approval, the results we have obtained provide preliminary evidence of safety and efficacy of a novel brain computer interface that is implanted without open brain surgery.

**Is there any new benefit for potential users of a BCI? Is there any improvement in terms of usability?**

**Nicholas:** The deliberate design features of the Stentrode™ Neural Interface System geared toward everyday use, provide many new benefits for the potential BCI users. As mentioned above, the implantation procedure does not involve invasive open-brain surgery or percutaneous connections, in turn, negating the associated health risks. The ITU is battery-less, meaning that time-domain data with full spectral information can be acquired without power consumption/device longevity issues and improve the likelihood of recording useful features.

The ETU that powers the ITU and acquires the data can be easily aligned by a lay person guided by an intuitive LED indicator. The custom-built software is entirely graphic-user-interface based, performs all necessary tasks to achieve BCI control, translates the decoded events into customizable HID standard inputs, operates in full performance on a laptop and does not require special programming knowledge for operation. Importantly, all components of the Stentrode™ Neural Interface System are portable and mountable onto a wheelchair and the entire system fits inside a small bag, which enables in-home and everyday use of the system.

**Is there any improvement in terms of speed of the system (e.g. bit/min)?**

**Nicholas:** Bench-top testing of the system with simulated signal with tentative settings (window size = 100 ms, temporal smoothing = 1 s boxcar, Fs = 2000 Hz) resulted in average decoding latency of between approximately 30–50 ms, thus, effective latency of ~150 ms. However, the actual information transfer rate of an actual user is unknown until the first participant of the clinical trial engages in user training.

**Is there any improvement in terms of accuracy of the system?**

**Nicholas:** We do not have performance data currently but expect to soon with the clinical trial. However, as mentioned above, the advantage of being able to record time-domain data with full spectral information (within the Nyquist limit—sampling rate of 2000 Hz) should favorably influence the decoder accuracy. Furthermore, because the Stentrode™ will be implanted in the Superior Sagittal Sinus (SSS) that is placed in between the left and right medial wall of the motor cortex, we are expecting to record robust and characterizable features upon a variety of mental strategies, which should further favorably improve the likelihood of achieving high accuracy.

**What are the next steps regarding your research?**

**Nicholas:** Our first in-human study is underway (Oxley et al. 2020), with results anticipated to be published and released publicly in the near future. Following extension of this trial to different countries and conditions, we believe that our technology will restore independence through direct brain-controlled communication to those in need.

# References

T. Oxley, N. Opie et al. Nat. Biotechnol (2016)
N.L. Opie, S.E. John et al. J. Neural Eng (2016)
N.L. Opie, S.E. John et al. Nat. Biomed. Eng (2018)
N.L. Opie, N.R. van der Nagel et al. IEEE Trans. Biomed. Eng (2017)
T.J. Oxley, P.E. Yoo et al. J. NeuroInterventional Surg (2020)

# Towards Brain-Machine Interface-Based Rehabilitation for Patients with Chronic Complete Paraplegia

**Solaiman Shokur**

**Abstract** We recently interviewed Solaiman Shokur about the project that his team submitted to the BCI Research Award in 2019. We then edited the interview and added images that Dr. Shokur kindly shared to provide more information about the team and project. Their project showed how a BMI-based protocol could provide partial neurological improvements for persons with spinal cord injuries. This is a very promising research direction, and several projects focused on improved rehabilitation therapy have been nominated for BCI Research Awards.

**Keywords** Brain-Machine Interface (BMI) · Spinal cord injury (SCI) · Neurorehabilitation · Treadmill · Multisensory integration

## 1 Introduction

In the first few decades of BMI research, most works aimed to help severely disabled patients by providing tools for communication (such as spelling) and/or control (such as a robotic arm). The prospect of using BMIs to help people recover motor function had been considered, but not well explored (Kuebler et al. 2001; Wolpaw et al. 2002). However, over the last decade, numerous papers have explored BMIs to support motor rehabilitation for people with stroke (e.g., Mrachacz-Kersting et al. 2014; Guger et al. 2018; Mane et al. 2020).

This chapter presents an interview with Dr. Solaiman Shokur about his work with the Walk Again Project team in São Paulo, Brasil, to extend this approach for

S. Shokur (✉)
Bertarelli Foundation Chair in Translational NeuroEngineering, Center for Neuroprosthetics and Institute of Bioengineering, Ecole Polytechnique Federale de Lausanne, Lausanne, Switzerland
e-mail: solaiman.shokur@epfl.ch

The BioRobotics Institute and Department of Excellence in Robotics and AI, Scuola Superiore Sant'Anna, Pisa, Italy

Neurorehabilitation Laboratory, Associação Alberto Santos Dumont para Apoio à Pesquisa (AASDAP), São Paulo, Brazil

C. Guger et al. (eds.), *Brain-Computer Interface Research*,
SpringerBriefs in Electrical and Computer Engineering,
https://doi.org/10.1007/978-3-030-60460-8_14

patients with chronic complete spinal cord injury (SCI). Recovering motor function is currently considered difficult or impossible for patients with some types of SCI, but Dr. Shokur describes how a new system integrating EEG-based BMI, with visuo-tactile feedback and locomotion could lead to new forms of treatment. Their project, team, and affiliation was (Fig. 1).

**The Walk Again Neurorehabilitation Protocol: A BMI-based Clinical Application to Induce Partial Neurological Recovery in Spinal Cord Injury Patients.**

Solaiman Shokur[1], Debora S. F. Campos[1], Ana R. C. Donati[1,2], Eduardo J. L. Alho[1], Mikhail Lebedev[3,4], Miguel Nicolelis[1,3,4,5,6,7,8,9].

[1]Neurorehabilitation Laboratory, Associação Alberto Santos Dumont para Apoio à Pesquisa (AASDAP), São Paulo, Brazil, 05440-000;

[2]Associação de Assistência à Criança Deficiente (AACD), São Paulo, Brazil, 04027-000;

[3]Department of Neurobiology, Duke University Medical Center, Durham, NC, USA 27710;

[4]Duke Center for Neuroengineering, Duke University, Durham, NC, USA 27710;

[5]Department of Biomedical Engineering, Duke University, Durham, NC, USA 27708;

[6]Department of Neurology, Duke University, Durham, NC, USA 27710;

[7]Department of Neurosurgery, Duke University, Durham, NC, USA 27710;

[8]Department of Psychology and Neuroscience, Duke University, Durham, NC, USA 27708;

[9]Edmond and Lily Safra International Institute of Neuroscience, Macaíba, Brazil;

Laboratory, Associação Alberto Santos Dumont para Apoio à Pesquisa (AASDAP), São Paulo, Brazil.

## 2   Interview

**What was the goal of your project?**

**Solaiman:** The goal was to study the neurological effects of BMI-based neurorehabilitation protocols for spinal cord injury (SCI) patients. The classical use of BMIs for SCI patients is as an assistive device. Simply said, it's a technique to bypass the lesion using a compensatory approach. We were interested to see how, under some conditions, it is possible to induce neurological recovery. We studied a neurorehabilitation protocol that integrated non-invasive (EEG-based) BMI with virtual reality and tactile feedback, with eight SCI patients with chronic lesion.

**How did you approach this goal?**

**Solaiman:** The current study was a follow-up to our work in 2016 when we observed improvements that, to our knowledge, had never been reported before to this extent in patients with severe motor injury (also referred to as motor-complete SCI patients; Donati et al. 2016; Shokur et al. 2018). In that study, we observed

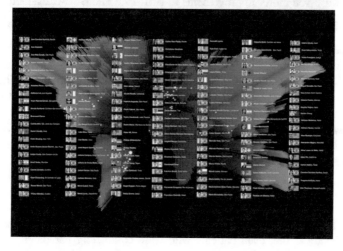

**Fig. 1** (Top) Dr. Solaiman Shokur, Senior scientist at the Walk Again Project, at the 2019 BCI Award Ceremony (right). The left and middle persons are Drs. Gunther Krausz and Brendan Allison. (Middle) Prof. Miguel Nicolelis, Principal Investigator of the Walk Again Project. (Bottom) The Walk Again Project consortium included researchers from 25 countries

significant motor and sensory recovery levels below the lesion of the patient. These were patients at the chronic phase of their lesion and had a complete loss of motor functions (some also sensory). After twelve months of training, they had recovered significant levels of sensation below the lesion and motor function in the lower limbs. Our goal for this second study was to isolate the effects of the non-invasive BMI to see if we could (a) reproduce our initial results and (b) investigate the impact of the BMI on top of locomotion training.

Our approach was to do a small clinical trial with a group of eight chronic complete SCI patients. We divided them into two subgroups. One group performed locomotion training only. The other group did the same locomotion training and had, additionally, one BMI session per week.

### Which technologies did you use?

**Solaiman:** We chose a purely noninvasive approach with EEG-based BMI, including event-related synchronization (mu rhythms in the motor cortex). We were looking specifically for leg motor imagery. We wanted to encourage our patients to imagine moving their legs—not imagining locomotion in an abstract way—and alternate between the left leg and right leg motor imagery. We used this decoding to move the corresponding leg of a 3D avatar in a virtual reality (VR) environment.

### The BMI was connected to a VR simulation of walking? Are you extending this to robotic devices?

**Solaiman:** Absolutely. We also did it with robotic devices and functional electrical stimulation (FES). The work is presented in another published paper from last year (Selfslagh et al. 2019), where we had BMI and FES, alone or in combination. We have also observed motor improvements in BMI-FES and BMI-exoskeleton.

### How did you use exoskeletons and treadmills?

**Solaiman:** In the study presented for the BCI Award, we used two modalities for locomotion training. We call it active locomotion training, as opposed to passive mobilization. In one paradigm, the patient was using a robotic gait trainer (Lokomat, Hokoma), and the physiotherapist was constantly motivating the patient to try to perform the task. So, the patient had an incentive to perform the task. The second locomotion paradigm was with body-weight supports on a rail (the ZeroG system, Aretech). Both subgroups had the exact same physical training and the same number of sessions (Fig. 2).

### What results did you get?

**Solaiman:** First, we observed some improvement in both motor and sensory functions for patients that followed the locomotion training alone. Therefore, the first conclusion is that even patients who were completely paraplegic and in the chronic stage of their lesion could benefit from an active locomotion training with the Lokomat and the body weight support. Second, importantly, we observed a systematically larger improvement for the group that followed both the locomotion and BMI training. When we reviewed their progress after five months, and again after

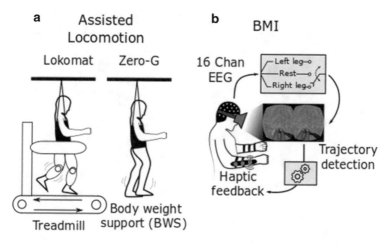

**a** Assisted Locomotion

Lokomat    Zero-G

Treadmill    Body weight support (BWS)

**b** BMI

16 Chan EEG

Left leg—o
Rest—o
Right leg—o

Haptic feedback

Trajectory detection

**Fig. 2** (A) The assisted locomotion training included training with a Lokomat and body weight support system (B) During the BMI task, the patient used left/right leg motor imagery to trigger the stepping of the corresponding leg of a 3D avatar seen in the first-person view

nine months, we saw that the BMI+locomotion group was always better than the locomotion-only group. This was specifically true in the motor domain. The biggest differences we saw were in the motor domain and proprioception.

**How many training sessions did you perform, and how long was each session?**

**Solaiman:** The patients came two times per week for approximately 30 weeks. The Locomotion-only group would do one day of Lokomat training and another day of body-weight support training. The BMI+Locomotion group would do the BMI the same day as the Lokomat (the BMI training was done right before the Lokomat), and then the body-weight training on another day. Therefore, both groups came twice per week.

The Lokomat and body-weight training lasted 45 min each. The BMI training was 4 times 6-minute runs.

**How important are these results for patients?**

**Solaiman:** Our result was quite important for demonstrating, for the first time, that AIS A or AIS B patients could recover neurological functions. To our knowledge, that had not been systematically shown to that extent. From a rehabilitation point of view, it is crucial to show that it is possible to improve those patients. The original paper had an important impact on the neurorehabilitation field. For clinicians, it's interesting because it's a relatively cheap technology. For hospitals or other clinical environments that already have Lokomats, adding a BMI is not very complicated. Even for the BMI itself, we are using well-known technologies, and you (Christoph) are engineering them and already using them for stroke patients. Since those techniques already exist and are even commercially available for stroke patients, they could be used in this protocol without changing too much. So, both the locomotion and BMI components

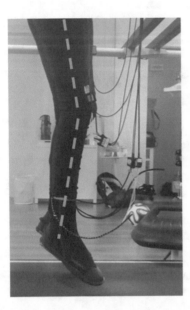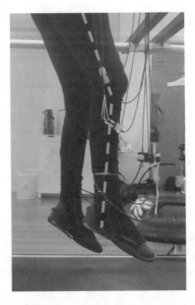

**Fig. 3** Patients who were initially diagnosed with complete loss of sensory-motor functions due to Spinal Cord Injury recovered significant motor function levels after training with the Walk Again Neurorehabilitation Protocol. Some of them could voluntarily contract their leg without external help (from Shokur et al. 2018)

are relatively easy, and the results were stronger than we expected and quite important for the field. However, the protocol is not yet optimized in time, and 30 weeks of training is certainly too long for this protocol to be deployed in an extended manner. We are currently working on optimizing the protocol and believe that intense training over a shorter time might induce the same recovery level, or maybe better (Figs. 3 and 4).

**How are your results applicable in clinical or real-life environments? Could you imagine this could be used in hospitals, rehab centers, or homes in a few years?**

**Solaiman:** All of the above. Our idea is to use BMIs in a neurorehabilitation protocol for SCI patients. Our protocol integrates BMI and locomotion training; we believe both aspects were essential to induce recovery. Indeed, other groups that have trained patients with the same kind of trauma with BMI alone did not observe this type of improvement. In the future, it might be possible to have the BMI part done at home and the locomotion part in a rehab center, but we have not tested it yet. So, it could be done to some extent at home.

**What are the next steps in your research?**

**Solaiman:** I think this first pilot-test was essential to show a proof of concept and reproduce our results from 2016; seeing the same effect in the second group

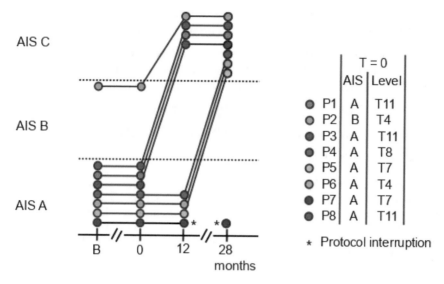

**Fig. 4** Prior work (Shokur et al. 2018) showed that training integrating non-invasive BMI, locomotion and visuotactile feedback induced significant recovery in a group of SCI patients. As a result, all seven patients that followed the protocol for 28 months improved to AIS C. Patient P7 voluntarily dropped out from the protocol after 12 months (personal reasons). Baseline measurement (B) was done by the clinical institution that followed the patients before they joined the protocol, and was done 1-3 years after the lesion. T = 0 stands for the first measurement done at the onset of the training, the 'Level', corresponds to the neurological level of the injury measured via with the ASIA test (see https://asia-spinalinjury.org/wp-content/uploads/2019/04/ASIA-ISCOS-IntlWo rksheet_2019.pdf)

was very interesting. Similar results have been reported in animal models, such as by Courtine and colleagues at EPFL (Bonizzato et al. 2018). We are interested in understanding the mechanism of what is going on because this is something missing at the moment. We have some hypotheses about why the patients improved to this extent. We are trying to understand this mechanism through fMRI protocols. One important step would be to understand what happens at the spinal cord level and the brain level and reproduce results with a larger group of patients and a sham BMI control group, which was not the case in our protocol.

### What is your experience in terms of the BMI performance of your spinal cord injury patients? Did they perform well?

**Solaiman:** Yes. The results were good. There are differences among patients. There were good performers and some average ones. We didn't have people who were completely at chance level. We didn't observe the effect that has been reported in the past called *BMI illiteracy*, which has been a major challenge for many years (Allison and Neuper 2010; Viduarre and Blankertz 2010; Thompson 2019). Maybe that's because the number of patients we had was small, but we didn't observe that. The people who were really good were so from the beginning. The other ones reached around 80% accuracy after a few sessions. We had six sessions at the beginning to check how they were performing, and then people stabilized around 75–80%.

**This is also our experience. In my understanding, a patient doesn't exist who is not able to control a BMI. Everybody can control it. Just people are doing something wrong if they don't reach an accuracy above chance level.**

**Solaiman:** That was our observation, too, absolutely.

**Thank you. That was a very nice explanation of what you did.**

**Solaiman:** Thank you

**Acknowledgements** This study was funded by the Brazilian Financing Agency for Studies and Projects (FINEP 01·12·0514·00), Brazilian Ministry of Science, Technology, Innovation, and Communication (MCTIC). We acknowledge the National Institute of Science and Technology (INCT) Brain Machine-Interface (INCEMAQ, Siconv 704134/2009) of the National Council for Scientific and Technological Development (CNPq), Brazilian Ministry of Science, Technology, Innovation and Communication (MCTIC).

# References

B.Z. Allison, C. Neuper, Could anyone use a BCI? in *Brain-Computer Interfaces* (Springer, London, 2010), pp. 35–54

M. Bonizzato, G. Pidpruzhnykova, J. DiGiovanna, P. Shkorbatova, N. Pavlova, S. Micera, G. Courtine, Brain-controlled modulation of spinal circuits improves recovery from spinal cord injury. Nature commun. **9**(1), 1–14 (2018)

A.R.C. Donati, S. Shokur, E. Morya, D.S.F. Campos, R.C. Moioli, C.M. Gitti, P.B. Augusto, S. Tripodi, C.G. Pires, G.A. Pereira, F.L. Brasil, Long-term training with a brain-machine interface-based gait protocol induces partial neurological recovery in paraplegic patients. Sci. Rep. 6, 30383 (2016)

C. Guger, J.D.R. Millán, D. Mattia, J. Ushiba, S.R. Soekadar, V. Prabhakaran, N. Mrachacz-Kersting, K. Kamada, B.Z. Allison, Brain-computer interfaces for stroke rehabilitation: summary of the 2016 BCI Meeting in Asilomar. Brain-Comput. Interfaces **5**(2–3), 41–57 (2018)

A. Kübler, B. Kotchoubey, J. Kaiser, J.R. Wolpaw, N. Birbaumer, Brain–computer communication: unlocking the locked in. Psychol. Bull. **127**(3), 358 (2001)

R. Mane, T. Chouhan, C. Guan, BCI for stroke rehabilitation: motor and beyond. J. Neural Eng. **17**(4), 041001 (2020)

N. Mrachacz-Kersting, N. Jiang, K. Dremstrup, D. Farina, A novel brain-computer interface for chronic stroke patients. In *Brain-Computer Interface Research* ( Springer, Berlin, Heidelberg, 2014), pp. 51–61

A. Selfslagh, S. Shokur, D.S. Campos, A.R. Donati, S. Almeida, S.Y. Yamauti, D.B. Coelho, M. Bouri, M.A. Nicolelis, Non-invasive, brain-controlled functional electrical stimulation for locomotion rehabilitation in individuals with paraplegia. Sci. Rep. **9**(1), 1–17

S. Shokur, A.R.C. Donati, D.S. F. Campos, C. Gitti, G. Bao, D. Fischer, S. Almeida, V.A.S. Braga, P. Augusto, C. Petty, E.J.L. Alho, Training with brain-machine interfaces, visuo-tactile feedback and assisted locomotion improves sensorimotor, visceral, and psychological signs in chronic paraplegic patients. PloS One **13**(11). https://doi.org/10.1371/journal.pone.0206464

M.C. Thompson, Critiquing the concept of BCI illiteracy. Sci. Eng. Ethics **25**(4), 1217–1233 (2019)

C. Vidaurre, B. Blankertz, Towards a cure for BCI illiteracy. Brain Topogr. **23**(2), 194–198 (2010)

J.R. Wolpaw, N. Birbaumer, D.J. McFarland, G. Pfurtscheller, T.M. Vaughan, Brain–computer interfaces for communication and control. Clin. Neurophysiol. **113**(6), 767–791 (2002)

# Recent Advances in Brain-Computer Interface Research: A Summary of the 2019 BCI Award and Online BCI Research Activities

**Christoph Guger, Michael Tangermann, and Brendan Z. Allison**

**Abstract** The introduction chapter of this book described the BCI Research Awards, selection criteria, nominees, and jury. Developing a good submission for a BCI Research Award is a formidable goal, and being nominated is even more demanding. This book has presented thirteen chapters by the authors of projects nominated for a BCI Research Award in 2019. Some of these chapters detailed the projects that were nominated, while other chapters comprised interviews with nominees. In this chapter, we review the 2019 BCI Research Awards Ceremony and present the winners. We also discuss emerging directions such as online BCI-related activities that have become much more prominent during 2020 due to COVID concerns.

**Keywords** Brain-computer Interface · EEG · ECoG · BCI Research Awards · BCI Foundation

## 1 The 2019 Awards Ceremony

The Awards ceremony for the 2019 BCI Research Award was part of the 8th Graz BCI Conference 2019 in Graz, Austria. This is consistent with prior years; we have typically announced the winners as part of a major international BCI conference. We announced the first, second, and third place winners at the Awards Ceremony at a rooftop venue at the Old University in Graz, with gorgeous views of the eastern Alps during late September. Figures 1 and 2 show that the dining, meeting, and balconies were filled with conference attendees eager to see the evening ceremony while enjoying local food and drinks.

C. Guger (✉)
g.tec medical engineering GmbH, Schiedlberg, Austria
e-mail: guger@gtec.at

M. Tangermann
Brain State Decoding Lab, Albert-Ludwigs-Universität Freiburg, Freiburg Im Breisgau, Germany

B. Z. Allison
Cognitive Science Department, University of California, San Diego, CA, USA

C. Guger et al. (eds.), *Brain-Computer Interface Research*,
SpringerBriefs in Electrical and Computer Engineering,
https://doi.org/10.1007/978-3-030-60460-8_15

143

**Fig. 1** The rooftop balcony area during the 2019 BCI Research Award Ceremony

Figure 3 shows Michael Tangermann introducing the BCI Award to the audience. Then, he and Gunther Krausz invited a representative from each of the nominated groups to join them on the stage. They announced the nominees, who received a certificate and other prizes, then remained onstage as Brendan Allison announced the winners.

## 2  The 2019 Winners

The 2019 BCI Research Award winners were:

**First Place Winner:**
**Decoding Speech from Intracortical Multielectrode Arrays in Dorsal Motor Cortex** (Fig. 4)
Sergey D. Stavisky[1], Francis R. Willett[1], Paymon Rezaii[1], Leigh R. Hochberg[2], Krishna V. Shenoy[1,3], Jaimie M. Henderson[1].
[1]Stanford University, USA.
[2]Brown University, Harvard Medical School, Massachusetts General Hospital, Providence VA Medical Center, USA.
[3]Howard Hughes Medical Institute, USA.

**Fig. 2** Jury member Selina Wriessnegger and other conference attendees during the BCI Award ceremony

**Second Place Winner:**
**BCI-based Neurofeedback Training for Quitting Smoking**
Junjie Bu[1], Kymberly D. Young[2], Wei Hong[1], Ru Ma[1], Hongwen Song[5], Ying Wang[1], Wei Zhang[1],
Michelle Hampson[3], Talma Hendler[4], Xiaochu Zhang[1,5].
[1]Hefei National Laboratory for Physical Sciences at the Microscale and School of Life Sciences, University of Science & Technology of China, Hefei, China.
[2]Department of Psychiatry, University of Pittsburgh School of Medicine, Pittsburgh, USA.

**Fig. 3** Michael Tangermann, Chair of the jury, presents the 2019 BCI Award

**Fig. 4** First place winners attend the Awards Ceremony from California via web

[3]Department of Radiology and Biomedical Imaging, Yale School of Medicine, New Haven, CT, USA.
[4]Functional Brain Center, Tel-Aviv University, Tel-Aviv, Israel.

**Fig. 5** The second place team receives their award from the jury and emcee

[5]School of Humanities & Social Science, University of Science & Technology of China, Hefei, China (Fig. 5).

**Third Place Winner:**

**Developing a Closed-loop Brain-computer Interface for Treatment of Neuropsychiatric Disorders using Electrical Brain Stimulation**

Yuxiao Yang[1], Omid G. Sani[1], Morgan B. Lee[2,3,4], Heather E. Dawes[2,3,4], Edward F. Chang[2,3,4], Maryam M. Shanechi[1,5].

[1]Ming Hsieh Department of Electrical Engineering, Viterbi School of Engineering, University of Southern California, USA.

[2]Department of Neurological Surgery, University of California, USA.

[3]Weill Institute for Neuroscience, University of California, San Francisco, USA.

[4]Kavli Institute for Fundamental Neuroscience, University of California, San Francisco, USA.

[5]Neuroscience Graduate Program, University of Southern California, USA.

Michael Tangermann concluded the ceremony by thanking the 2019 jury and the conference chair, Prof. Gernot Müller-Putz, shown in Fig. 6. Prof. Gernot Müller-Putz is the Head of the Institute of Neural Engineering and the Laboratory of Brain-Computer Interfaces at the Graz University of Technology. This institute has hosted several BCI Conferences in Graz.

# 3    Conclusion and Future Directions

We have always concluded our annual BCI Research Award books with a discussion chapter that presents the winners and explores emerging directions. In the discussion chapter from last year's book (Guger et al. 2020) other recent work, we addressed

**Fig. 6** Prof. Gernot Müller-Putz thanks the attendees of the awards ceremony

the rise of activities to help inform and educate people about BCIs, especially people new to the field. Many of these activities were public, hands-on events such as BCI hackathons, conferences, Cybathlons, and live, real-time BCI competitions (Vala-jame et al. 2017; Guger et al. 2019; Ortiz et al. 2019; Allison et al. 2020). Other groups have also published work about such activities (Statthaler et al. 2017; Novak et al. 2018; Perdikis et al. 2017, 2018; Lotte et al. 2019). Since then, concerns with COVID and future pandemics has made live activities with many people unfeasible. How can the BCI community adapt?

As noted in the introduction to this book, some adaptations to COVID have already been successful, such as online conferences, workshops, and training opportunities. Different organizers have told us their events had hundreds or thousands of online attendees, with strong interest in future online events. Some other public BCI activities, such as Virtual Users' Fora or offline BCI data analysis competitions, have not required in-person participation (Blankertz et al. 2006; Huggins and Wolpaw 2014; Peters et al. 2015; Huggins et al. 2017, 2019).

Other events may substitute for events like in-person BCI hackathons and inter-active demonstrations. For example, online courses could include exercises that students could perform at home with BCIs, which is becoming more feasible as BCIs become cheaper and easier to use. Webinars in which hosts demonstrate different aspects of hands-on BCI use, including time for questions from attendees, have been effective for many years. While we hope that concerns about pandemics abate and in-person activities become prevalent again soon, these activities are not only replace-ments for events that have been cancelled or postponed. Online events focused on

BCIs or other activities will become more prominent even after a possible "return to normalcy."

Organizing and executing high-quality BCI activities is time-consuming and difficult, even without the challenges of managing in-person interaction. Booking guest speakers, planning, scheduling, advertising, and executing BCI events is not for the timid nor overconfident. Aside from common challenges like a good internet connection and software to handle presentations and questions effectively, the audience may have a wide range of BCI backgrounds. Thus, some material may be too simple or advanced, and discussions may flounder amidst a confused audience. This challenge might be offset by announcing the difficulty level of specific talks or the entire event to help people decide whether they wish to participate and prepare accordingly.

We've made some changes to the BCI Research Awards and book series since 2010. The BCI Research Awards began with only ten nominees and one winner, and now has twelve nominees and first, second, and third place, winners. We explored special chapters that address major topics in BCIs or other noteworthy BCI projects. We recently added interviews with winners. The interviews with winners were well-received in last year's book, and we plan to continue these interviews in future books.

We may expand the BCI Research Award and books in different ways. The annual Awards Ceremony could aim to attract even more attendees, with more publicity. We are considering online or in-person discussion panels with nominees, perhaps in addition to the Awards Ceremony. Instead of simply summarizing trends in discussion chapters, we could instead develop a paper with a ten-year retrospective analyzing trends across all awards. We are still open to different chapters that would be consistent with other themes in each book and interest our readers.

Dr. Tangermann, the Chair of the 2019 Jury, said: "The strong growth BCI-related publications in the recent years shows its attractiveness to the community, but comes with the drawback that it became harder to follow this rapid development. The BCI Award provides a remedy to this problem, as highly attractive novel projects are carefully selected by an international committee and highlighted to the community." Like previous years, this year's BCI Research Award and book aimed to recognize and encourage the top projects in our field. Some chapters could provide background information, connections, or inspiration for readers who are new to the field as well as experts doing related work. You might have ideas for your own BCI project. We might someday give you a certificate for being nominated or even winning a BCI Award and/or cite your work in our books or other papers. Despite the uncertainty with pandemics, we still expect an ongoing increase in BCI activity in the near future and beyond. BCI systems have many challenges to overcome for practical use with healthy people and varied patient groups, and hence new ideas and applications can have a very strong impact. The field needs people from various disciplines including neuroscience, programming, mathematics, medicine, ethics, human-computer interaction, and many engineering domains including hardware, mechanical, electrical, and biomedical. BCIs for gaming, music art, robotics, or other goals require specialists in relevant fields as well. Healthy users and patients can contribute as volunteers or paid participants to help many BCI labs develop and test systems. Hence, whatever your background or interests, you might be able to contribute to BCI research and development. Thank you for reading this book, which we hope has been informative and helpful.

## Works Cited

B.Z. Allison, A. Kuebler, J. Jin, 30 + Years of P300 BCIs. Psychophysiology **57**(7) (2020). https://doi.org/10.1111/psyp.13569

B. Blankertz, K.R. Muller, D.J. Krusienski, G. Schalk, J.R. Wolpaw, A. Schlogl, G. Pfurtscheller, J.R. Millan, M. Schroder, N. Birbaumer, The BCI competition III: Validating alternative approaches to actual BCI problems. IEEE Trans. Neural Syst. Rehabil. Eng. **14**(2), 153–159 (2006)

C. Guger, B.Z. Allison, M. Walchshofer, S. Breinbauer, The BR4IN. IO Hackathons, in *Brain Art* (Springer, Cham, 2019), pp. 447–473

C. Guger, B.Z. Allison, K. Miller, Highlights and Interviews with Winners, in *Brain–Computer Interface Research* (Springer, Cham, 2020), pp. 107–121

J.E. Huggins, J.R. Wolpaw, Papers from the fifth international brain–computer interface meeting. J. Neural Eng. **11**(3), 030301 (2014)

J.E. Huggins, G. Müller-Putz, JR Wolpaw, (2017). The sixth international brain–computer interface meeting: advances in basic and clinical research. Brain comput. interfaces (Abingdon, UK), **4**(1–2), 1

J.E. Huggins, C. Guger, E. Aarnoutse, B. Allison, C.W. Anderson, S. Bedrick, W. Besio, R. Chavarriaga, J.L. Collinger, A.H. Do, C. Herff, Workshops of the seventh international brain-computer interface meeting: not getting lost in translation. Brain-Comput. Interfaces **6**(3), 71–101 (2019)

F. Lotte, M. Clerc, A. Appriou, A. Audino, C. Benaroch, P. Giacalone, C. Jeunet, J. Mladenović, T. Monseigne, T. Papadopoulo, L. Pillette, Inria Research & Development for the Cybathlon BCI series. In International Graz Brain-Computer Interface conference (2019, September)

D. Novak, R. Sigrist, N.J. Gerig, D. Wyss, R. Bauer, U. Götz, R. Riener, Benchmarking brain-computer interfaces outside the laboratory: The Cybathlon 2016. Front. Neurosci **11**, 756 (2018)

M. Ortiz, E. Iáñez, C. Guger, J.M. Azorín, The art, science, and engineering of BCI Hackathons. In *Mobile Brain-Body Imaging and the Neuroscience of Art, Innovation and Creativity* (Springer, Cham, 2019), pp. 147–155

S. Perdikis, L. Tonin, J.D.R. Millan, Brain racers. IEEE Spectr. **54**(9), 44–51 (2017)

S. Perdikis, L. Tonin, S. Saeedi, C. Schneider, J.D.R. Millán, The Cybathlon BCI race: successful longitudinal mutual learning with two tetraplegic users. PLoS Biol. **16**(5), e2003787 (2018)

B. Peters, G. Bieker, S.M. Heckman, J.E. Huggins, C. Wolf, D. Zeitlin, M. Fried-Oken, Brain-computer interface users speak up: the Virtual Users' Forum at the 2013 International Brain-Computer Interface meeting. Arch. Phys. Med. Rehabil. **96**(3), S33–S37 (2015)

K. Statthaler, A. Schwarz, D. Steyrl, R. Kobler, M.K. Höller, J. Brandstetter, L. Hehenberger, M. Bigga, G. Müller-Putz, Cybathlon experiences of the Graz BCI racing team Mirage91 in the brain-computer interface discipline. J. Neuroeng. Rehabil. **14**(1), 1–16 (2017)

A. Valjamae, L. Evers, B.Z. Allison, J. Ongering, A. Riccio, I. Igardi, D. Lamas, The Brain-Hack project: exploring art-BCI hackathons. In Proceedings of the 2017 ACM Workshop on An Application-oriented Approach to BCI out of the laboratory (2017, March), pp. 21–24. https://doi.org/10.1145/3038439.3038444